GREAT NORTHERN RAILWAY
ORE DOCKS OF LAKE SUPERIOR
PHOTO ARCHIVE

Douglass D. Addison, Sr.

Iconografix
Photo Archive Series

Iconografix
1830A Hanley Road
Hudson, Wisconsin 54016 USA

Library of Congress Card Number: 2002104765

ISBN-13: 978-1-58388-073-9
ISBN-10: 1-58388-073-9

Reprinted May 2012

Printed in The United States of America

Cover and book design by Shawn Glidden

Copyediting by Suzie Helberg

COVER PHOTO: See page 116.

BOOK PROPOSALS

Iconografix is a publishing company specializing in books for transportation enthusiasts. We publish in a number of different areas, including Automobiles, Auto Racing, Buses, Construction Equipment, Emergency Equipment, Farming Equipment, Railroads & Trucks. The Iconografix imprint is constantly growing and expanding into new subject areas.

Authors, editors, and knowledgeable enthusiasts in the field of transportation history are invited to contact the Editorial Department at Iconografix, Inc., 1830A Hanley Road, Hudson, WI 54016.

Table of Contents

DEDICATION

This book is dedicated to my wife and best friend, Lynne, and to thank the Lord for the blessings He gives each day.

ACKNOWLEDGMENTS

In 1968, I lived in Willmar, Minnesota and was introduced to Lee Thompson, an ex-Great Northern employee. Lee is into model railroading the GN and soon after I met him, I was too. I became addicted to the GN on the Mesabi Iron Range and the four GN ore docks at Allouez, Wisconsin.

In 1982, I joined the Great Northern Railway Historical Society and their Reference Sheet 65 (Great Northern Iron Range) started my interest in this fantastic, unique industry. Stuart Holmquist was the contributor to that reference sheet. Stuart has been very helpful in providing me with pictures, information and advice.

The Burlington Northern gave me permission to photograph the two iron ore docks still in operation at that time (Ore Docks One and Two). I took over 500 pictures in four trips. The Burlington Northern also gave me the blueprints that I used to draw the drawings in this book, except Ore Dock Three (part plan showing pocket framing and track construction), courtesy Great Northern Railway Historical Society. The drawings of the maintenance building and interior were tape measured by Lee Thompson and myself, and I etched the drawings.

The drawings were done on 12x18-inch papers and took 3 1/2 years to complete. They were again reduced to fit the pages of this book. Copies of the drawings can be obtained by writing the author at: 1381 105th Ave. NW, Coon Rapids, MN 55433.

A special thank you to Ms. Suzanne L. Burris, Curator of the archives at the Burlington Northern Santa Fe Railroad, who helped receive permission to use the pictures and information used for most of this book.

I started out to build an iron ore dock on my model railroad and ended up with a mission to document the operations of the largest iron ore docks in the world.

Air view of the Allouez yard with ore docks in background. The yard had a capacity of over 4,000 cars.
Burlington Northern Santa Fe Railway

INTRODUCTION

The following article is reprinted with permission from the Burlington Northern Santa Fe Railroad. It is from the April 1925 Great Northern **Semaphore**. By 1928, Ore Docks One and Two were converted from wood to concrete and cement. This left Ore Dock Three; the only Great Northern ore dock that would remain wood in construction. The changes made from around 1925-onward are shown later.

GREAT NORTHERN ORE DOCKS

Description of the Four Great Docks at Allouez (Superior), Wis., Where Millions of Tons of Ore are Transferred Each Season from Cars to Boats

By H. J. Seyton, District Engineer.

[This article describes Docks One, Two, Three and Four owned by]...the Great Northern Railway company at Allouez, just inside the harbor entrance from Lake Superior, where the ore brought down from the Mesabi range is transferred by gravity from hopper-bottomed ore cars to boats for shipment down the lakes. Each dock has four tracks arranged in pairs over a set of pockets on each side of the dock, to provide for loading the pockets to full capacity without tripping. The pockets are on twelve foot centers, and the cars, twenty-four feet long, are spotted over alternate pockets. The entire scheme of ore handling is carried out on this basis.

Dock No. 1, pioneer unit, was constructed of timber in 1892 and rebuilt of the same material in 1907. It extends out to the harbor line, is 2,244 feet long, and has 374 pockets of 300 tons capacity each, or a total of 112,000 gross tons. The dock is 73 feet above water level and 63 feet wide. One-third of this dock has been dismantled and is now being rebuilt of steel and concrete. The outer two-thirds of the dock will be kept in service this season by the use of a temporary trestle which permits construction work on the new portion to proceed without interruption of

the ore traffic. The dock, when completed, will be similar to dock No. 2, electrified throughout, whereas now it is manually operated.

Dock No. 2, our newest unit, is the most modern dock on the Great Lakes, being of steel and concrete and all electrically operated. The original dock, a timber structure erected in 1899, was replaced by the present one in 1923. It extends to the harbor line, is 2,100 feet long, 56 feet wide, and 80 1/2 feet above water level, or 7 1/2 feet higher than the old dock. This additional height enabled us to increase the pocket capacity fifty tons, giving 350 pockets of 350 tons capacity each, or a total of 122,500 tons. The foundation of the dock consists of a concrete mattress placed at water level on 16,534 plumb and 1,422 batter piles, the latter being required for side thrust resistance. The substructure contains 40,000 cubic yards of reinforced concrete and 770 tons of reinforcing steel. The superstructure has 11,490 tons of steel construction; the pocket partitions are of concrete slabs.

Dock No. 3 was originally constructed in 1902 with 160 pockets and increased to 326 pockets in 1906. The first portion was rebuilt in 1917 and the outer half in 1921. The dock extends to harbor line, is 1,956 feet long, 60 feet wide, and 77 feet above water level. Its 326 pockets of 300 tons capacity give it a total capacity of 97,800 tons. This dock, while of timber construction, is electrically operated, and is the most modern wooden ore dock on the lakes. With the completion of No. 1 dock, it will be our only remaining wooden one. Most timber docks have maple pocket linings, but this dock has precast concrete slabs for pocket linings, these having proved to be the most efficient and economical as a sliding surface in handling ore.

Dock No. 4, built in 1911, is our first permanent ore dock. It is of Toltz design, having barrel shaped steel pocket fronts, crib foundation, reinforced concrete substructure, and steel superstructure. The dock extends to harbor line, is 1,812 feet long, 62 feet wide and 75 feet above water level. It has 302 pockets of 300 tons capacity each, making a total of 90,600 tons.

Our four ore docks have a combined capacity of 423,100 tons, equivalent to about 55 train or 47 boat loads.

A fireproof electrical power house, using purchased energy, was constructed on the shore end of the matters when rebuilt in 1923. This contains generators for operating spout hoists on docks two, three and four, a compressor to furnish air for dumping hammers, a steam plant for heating office and eating house, and the water pumps. The pumps furnish water for the dock and also for the engine terminal some two and one-half miles distant. During the ore season we use, daily, 800,000 gallons of water and the only cost for this vast quantity is the electricity for pumping it, water being obtained form the bay through an intake in the floor of the power house.

The dock office is located on a trestle on the shore between docks two and three, level with the dock decks. Cross walks connect all four docks, affording easy access for employees, or from one to another. An eating house for dock employees is located on the approach to each dock. [Note: This was replaced in 1945 by two-story brick General Service Building and Maintenance Building described on pages 71-81.]

The slips for boats between docks are 200 feet wide and dredged 22 feet deep. This width will permit a large boat to pass between two others tied up at the dock on each side of the slip.

Ore cars from the Allouez yard are brought up on an elevated steel approach 2,892 feet long leading to dock No. 2, with diverging approaches to the other docks. The track from the yard to the steel approach has an ascending grade of .8 percent and grade on the approach proper varies from .7 to 1 percent.

All dock repair work is done during the winter, principally because there is no traffic, but also to take advantage of working on the ice in the slips, which method is much cheaper than performing work with floating equipment.

LOADING OF IRON ORE ON THE MESABI RANGE

By Geo. R. Clarke, Chief Dispatcher, Kelly Lake, Minn.

The discovery of iron ore on the Mesabi range was made in the fall of 1890 adjacent to the present Mesabi mountain mine near Virginia by the Merritt brothers of Duluth. This range is approximately seventy-five miles northwest of Duluth and extends from southwest to northeast with a productive length of about eighty miles. During the year about ten thousand active workers are employed in and about the mines. About 65 percent for the ore moved down the Great Lakes comes from this range and the State of Minnesota produces a like percentage of the total ore mined in the United States.

The Great Northern Railway company has been moving iron ore from the start of the Mesabi range production. The first shipment was 4,245 gross tons in 1892 over the old Duluth and Winnipeg railway, and our peak season was 1923 with a movement of 15,725,551 gross tons. [Note: As this article was printed in 1925, the real peak season was in 1953 when 32,330,722 tons were shipped.]

With the first discovery and mining of iron ore on the Mesabi range, the wonderful business wisdom of the late James J. Hill was displayed by the acquisition of a large number of mineral leases on lands adjoining the railroad. These leases are now held by the Great Northern Iron Ore properties. Many of the properties are being worked by mining companies under appropriate agreements on a royalty basis.

Ore was first mined by the underground method, being raised through a shaft to the surface, and this method is still used where conditions require. The underground mines hoist throughout the year, loading direct into cars during the shipping season and stockpiling during the winter. These stockpiles are loaded out with steam shovels during the shipping season. The railroad furnishes empty cars and the mines move them by gravity for loading at the shafts. For the stockpiles, we furnish cars with locomotive service in manner similar to our own steam shovel operation at gravel pits.

Later a more economical operation than shaft mining was developed for recovering the ore which is the "open pit." In this process the overburden is stripped off and the ore mined direct with steam shovels. It is now generally considered that open pit mining is the more economical where there is not more than two feet of overburden to one foot of ore, the total thickness of the ore body and the area of the deposit considered. For this service we furnish cars in yards on the surface adjacent to the pit and the mining companies, with their own locomotives, handle cars in and out of pit.

As the ore mining progressed and more attention was given to the grading of ore, it was found that some of the ore did not have sufficient iron content; that some was too wet, chunky, etc. This resulted in the establishment of beneficiation plants—washers, dryers, crushers and screening plants. For these plants we furnish cars for loading in the same manner as at the shafts.

Most of the mines have large modern pumping plants for dewatering pits as well as underground workings. At the various mines there are "locations," or villages of company houses, where employees live adjacent to their work.

Distribution of cars to the various mines is made through the chief dispatcher's office at Kelly Lake, also the assembling of loads and trains is handled there until turned over to the Superior office on reaching the main line at Brookston, Gunn or Swan River. There are about thirty transfer crews working during the busy season, delivering the empties and bringing the loads into Kelly Lake, where they are assembled into trains for Allouez.

After the cars are loaded at the mines, a sample of ore is taken from each car. This sample consists of about twenty small scoops of ore and is put through a laboratory analysis to determine the content of silica, iron, phosphorus, moisture, etc., to establish the grade. When trains depart from Kelly Lake a grade report is promptly given the mining companies covering the individual cars in the train loaded with the ore, and usually the grade of the ore is furnished the dock before the train arrives there.

TRANSPORTATION OF ORE
By R. E. Kelly, General Superintendent's Office, Duluth, Minn.

While the Range employees distribute the empties, assemble the loads and get the trains out of Kelly Lake and the dock employees dump the cars and load the boats, the general superintendent's office "fill the gap"—sees that the proper ore is moved at the right time.

A check is made by this office with the dock each morning, to ascertain the boat line-up and the grades of ore available for the different cargoes; with the Range as to the number of and time locomotives and crews will be available and the location of loads from the various mines; of the ore in transit, which is furnished by telegraph from each train conductor as he leaves Kelly Lake, specifying the number of loads in his train from the various mines. From these checks is determined the quantity and kind of ore needed to load the boats named to us, in the order in which they are due; and instructions are given the Range on the movement of the ore for the next twenty-four hour period, or longer, accordingly. If there are shortages of ore for individual cargoes that must be immediately secured for boats close by, we get in touch with the graders at the mines to secure detailed information as to waybills outstanding or request a substitution of other waybills in order to expedite assembling of cargoes and insure the boats' dispatch. Individual waybills are then located on the range and those loads moved in preference.

It is of interest in connection with this account of ore transportation that Engineman John Arten, on the Mesabi run is a full blood Chippewa Indian, chief of the Fond du Lac band of that tribe. Arten left the reservation at an early age, and after several years of wandering, was picked off the roads of a South Shore passenger train near our Allouez round house in June, 1903. The young Indian got a job in the roundhouse, and by application to duty and the help which he, himself, says he received form Great Northern officials, worked up in his profession until, on June 23, 1923, extra 2022, he hauled the biggest train of iron ore probably ever hauled, 13,221 tons, 103 miles in seven hours and eighteen minutes. There were 9,095 long tons of iron ore in the train, which, if smelted into arrowheads, would have pointed all the arrows ever shot by the Red Man form time immemorial.

We are connected with the Range by telegraph and telephone, and by keeping in close touch with the ore as it is assembled or moves down in trains, we are always in a position to say what each train will have and when it will arrive at Allouez, and can arrange for expeditious handling on arrival, where it is necessary. This method of handling has proved very satisfactory to the mining and boat companies, and to ourselves as well.

We keep constant watch of the tonnage carried on hand by the mining companies and "follow up" with the boat companies to furnish vessels. When they cannot be provided promptly, it is necessary to reduce or shut off the car supply at the mines to prevent certain mines loading excess ore, which would cause a shortage of cars and hamper the operations of other companies. When business is heavy this office supervises the distribution of cars by the dock agent to see that the various mining companies are taken care of in proper proportion to their output and boat supply.

There is a certain fascination about railroad work, and this, coupled with lake transportation, makes an interesting combination such as we have nowhere else on the railroad. Lake movement of ore started back in the 90's with sailing vessels, followed with steam boats. The latest addition came last year with the appearance of two electrically driven boats with oil burning Deisel engines (opposed piston type) making the power. A comparison of the pictures shown will illustrate the evolution of boats as well as docks. Lake boats are "U" shaped or flat bottomed compared to "V" shaped or sharp keel boats used on salt water. They range in size from 400 to better than 600 feet long, with a width of 50 to 64 feet, and a rated carrying capacity of from 5,000 to 13,000 gross tons. The usual draft is 19 1/2 to 20 feet, governed by the depth of water at Soo locks. The record ore movement from Duluth-Superior harbor was in 1923, when 42,350,918 net tons were shipped. The record cargo in 1924 was 14,004 net

tons of ore. The season of navigation is usually a week or two earlier and later.

The vessel companies' headquarters are chiefly in Cleveland, but the larger companies have vessel agents in Duluth, who handle the boats between here and the Soo, in the same way we divide our railroad by divisions and handle trains with different sets of dispatchers. The location of the general superintendent's office in Duluth, in close contact with the vessel and mining departments of the various companies, has resulted in a co-ordination of the customers' departments with the railroad that has greatly improved our handling of the business. We strive to secure vessel tonnage for the ore promptly when mined and to eliminate complaints on car supply by seeing that cars are not loaded by the various companies out of proportion to the boat tonnage furnished by them.

The whole question of loading and moving the ore from the Mesabi range is governed by the boat supply, and this in turn is restricted to what can be handled at the lower lake ports, either for storage or direct shipment to the furnaces.

HANDLING OF ORE ON DOCKS
By A. W. Elmgreen, Agent, Allouez Dock

On arrival of an ore train at Allouez, the weighmaster makes a list of cars of each grade, or block, showing position in train, and copies this list to the yardmaster and the dock office. Because of wide variation in the quality of the ore and the large number of companies served by us, there are more than fifty shipping grades handled. These grades are subdivided, and during a busy season there is a daily average of about seventy-five subdivision of ore to be switched out and dumped separately.

The classification of ore at the yards is under the supervision of the yardmaster. In addition to the receiving and "empty" yards, we have at Allouez two classification yards containing working space for 1,500 cars. This is now being increased to 2,700 cars capacity. Trains are shoved from the receiving yard over the scales and dropped off the "hump" onto classification tracks according to grade. As cars pass over the scale they are automatically weighed and cards are put on them indicating the grade of the ore contained.

Orders to ship ore in vessels are received from the boat or shipping companies, and the mining company graders then specify what particular ore is to be used for the cargo. The dock agent gives the necessary orders to the general foreman, who, as space in the dock is available, orders from the yard such ore as will fit their requirements. The aim is to use available dock space for cargoes in the order the boats are due. Ore is shoved to the dock in "cuts" of about thirty cars and the switch foreman delivers to the dock foreman a slip indicating the number of cars and grades in his cut. He in turn receives orders showing the pockets over which to spot the cars. The slip is turned over to the check clerk who checks the cars and pockets and also compares the slip with tags on the cars to see that they properly correspond.

Dumpers work in two-man gangs, preceded by trappers, who open the cars, followed by cleaners who clean them and then by trappers who close the hoppers. While much of the ore runs out freely when cars are opened, the structure of some is such that it sticks, or "hangs" in the car, and this necessitates considerable work to dump it. Ordinarily the only tool required by the dumpers is a six foot steel bar pointed at one end. First they pry the ore out of the hopper from the bottom, then pound the side of the car with the blunt end of the bar until the main body of the load drops out. They then climb on the car and poke out the balance. During recent years the unloading of tough ore with air hammers has been introduced and worked successfully. Washed ore is nearly always "tough," and particularly so if it has stood under load some days.

The first thousand new cars of large capacity were received in 1920, and with an additional 500 to be delivered this spring, we will have 3,250. These cars have given a wonderful account of themselves in carrying capacity and fast dumping.

The boat loading is under the supervision of the dock foreman. The mate of the vessel is on duty continuously during the loading and gives all orders concerning delivery of ore in different hatches, etc. The load, in boats, is trimmed by raising or lowering the spouts. Boat hatches are spaced 12 or 24 feet apart, fitting in with ore dock pocket construction, and several pockets can be run at one time. Some ores slide out very readily when pocket doors are opened; others require considerable poking at the door opening and poling from on top to get the pockets clean.

After boats leave the dock they run down Lake Superior, reaching the Soo locks in about 36 hours, according to the speed, and then on down to the lower lake port where the cargo is unloaded. Boats in a regular trade throughout the season make the round trip in a week.

When the cargo is broached at destination samples are again taken to check the analysis made by the shippers. Samples are always taken after the cargo is broken so that no top surfaces are taken where the ore may have been rained on or dried out.

Lower lake docks are variously equipped for unloading ore from boats by suspending a clam to drop in the hatches and carry the ore back into cars or on dock. The most noteworthy machines are electrically operated Huletts with 10 to 17 ton grab buckets on a rigid arm and tilting frame, all on rails, as shown in the picture. The operator rides just above the bucket and controls the entire machine, first going down, getting a load and then swinging up and back, where ore is dropped into a weighing hopper, from which it is discharged into cars underneath or to loading pit, from which it is handled by dock bridge out onto the dock stockpile. The cars are moved inland to furnaces where they are dumped by car dumpers and the ore stockpiled for future use.

Great Northern Allouez dock complex, eastern tip of Duluth-Superior Harbor entrance showing ore docks (from left to right) Four, Three, Two and One. Northern Pacific Ore Dock is seen to the right. ***Great Northern Railway Historical Society***

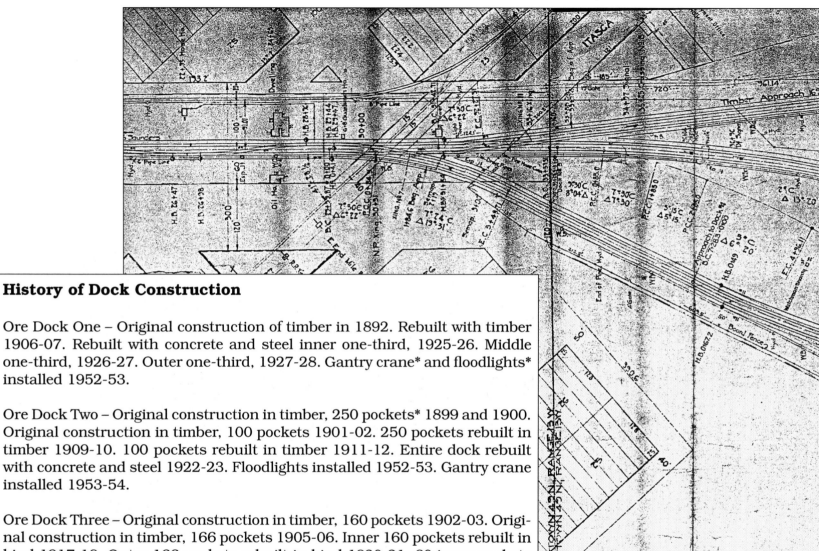

History of Dock Construction

Ore Dock One – Original construction of timber in 1892. Rebuilt with timber 1906-07. Rebuilt with concrete and steel inner one-third, 1925-26. Middle one-third, 1926-27. Outer one-third, 1927-28. Gantry crane* and floodlights* installed 1952-53.

Ore Dock Two – Original construction in timber, 250 pockets* 1899 and 1900. Original construction in timber, 100 pockets 1901-02. 250 pockets rebuilt in timber 1909-10. 100 pockets rebuilt in timber 1911-12. Entire dock rebuilt with concrete and steel 1922-23. Floodlights installed 1952-53. Gantry crane installed 1953-54.

Ore Dock Three – Original construction in timber, 160 pockets 1902-03. Original construction in timber, 166 pockets 1905-06. Inner 160 pockets rebuilt in kind 1917-18. Outer 166 pockets rebuilt in kind 1920-21. 60 inner pockets rebuilt in spring 1942. Next 86 pockets rebuilt spring of 1944. 80 pockets rebuilt 1952-53.

Ore Dock Four – Constructed with concrete and steel in 1911.

*Terms may be unfamiliar and are described in detail in later chapters.

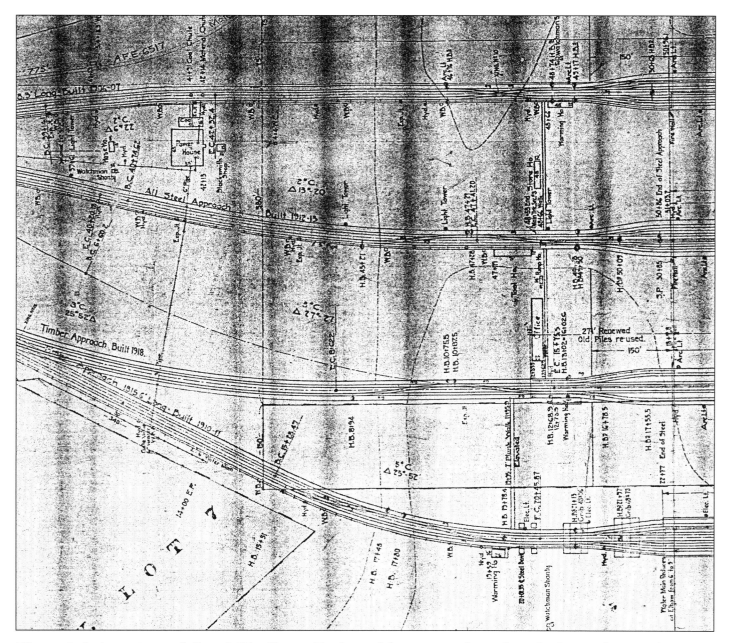

Blueprint: top view of Great Northern Railway blueprint of steel rails from the yard to approaches and onto all four ore docks (full overview of docks shown on pages 12-15). *Great Northern Railway, courtesy Burlington Northern Santa Fe Railway*

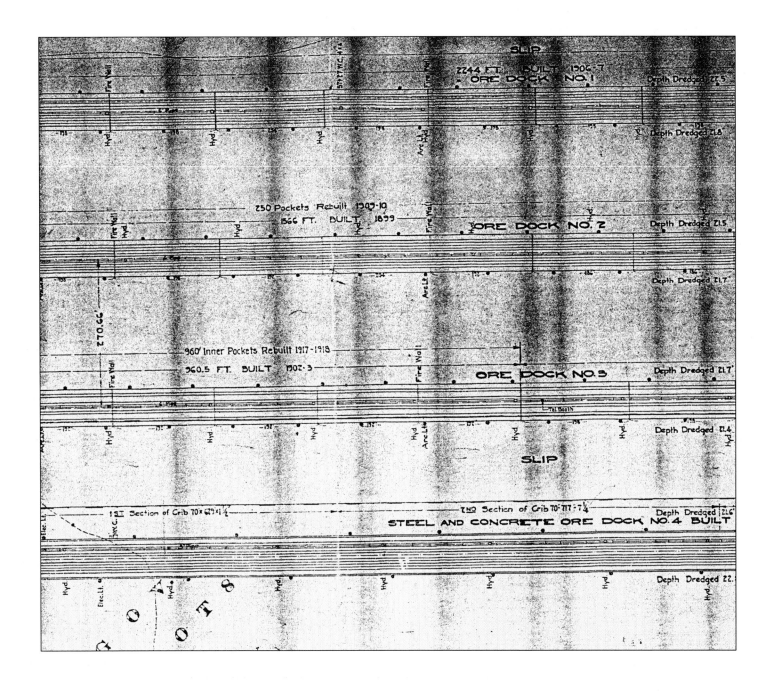

SLIP

2244 FT.   BUILT 1906-7

ORE DOCK NO. 1   Depth Dredged 25.5

Depth Dredged 21.8

250 Pockets Rebuilt 1909-10

1566 FT. BUILT 1899

ORE DOCK NO. 2   Depth Dredged 21.5

Depth Dredged 21.7

960' Inner Pockets Rebuilt 1917-1918

960.5 FT. BUILT 1902-3

ORE DOCK NO. 3   Depth Dredged 21.7

Depth Dredged 21.4

SLIP

1ST Section of Crib 70 x 69 x 1½   2ND Section of Crib 70-717-7¼   Depth Dredged 21.6

STEEL AND CONCRETE ORE DOCK NO. 4 BUILT

Depth Dredged 22.

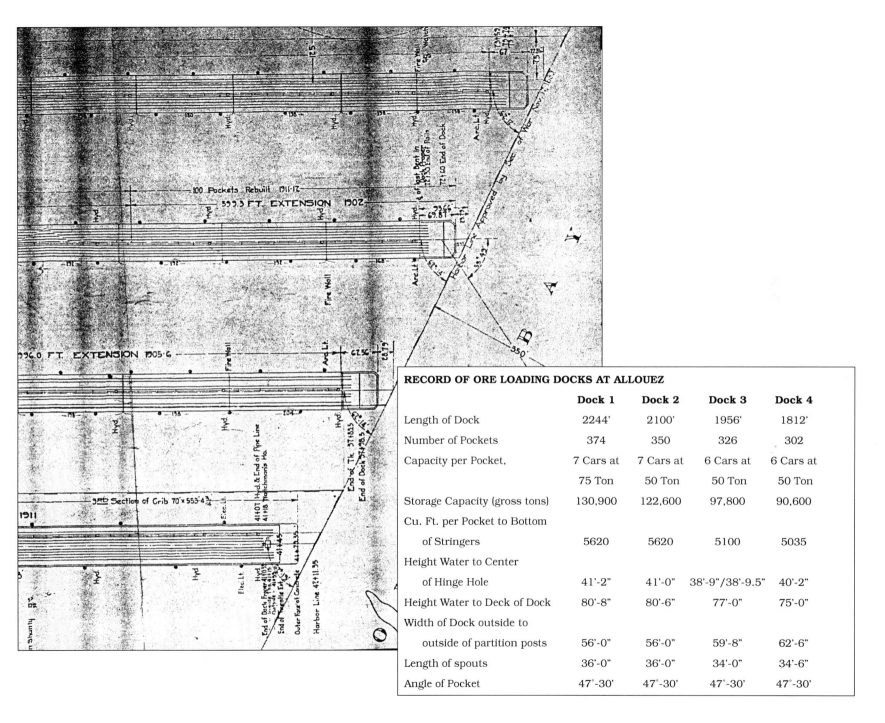

RECORD OF ORE LOADING DOCKS AT ALLOUEZ				
	Dock 1	**Dock 2**	**Dock 3**	**Dock 4**
Length of Dock	2244'	2100'	1956'	1812'
Number of Pockets	374	350	326	302
Capacity per Pocket,	7 Cars at	7 Cars at	6 Cars at	6 Cars at
	75 Ton	50 Ton	50 Ton	50 Ton
Storage Capacity (gross tons)	130,900	122,600	97,800	90,600
Cu. Ft. per Pocket to Bottom				
of Stringers	5620	5620	5100	5035
Height Water to Center				
of Hinge Hole	41'-2"	41'-0"	38'-9"/38'-9.5"	40'-2"
Height Water to Deck of Dock	80'-8"	80'-6"	77'-0"	75'-0"
Width of Dock outside to				
outside of partition posts	56'-0"	56'-0"	59'-8"	62'-6"
Length of spouts	36'-0"	36'-0"	34'-0"	34'-6"
Angle of Pocket	47°-30'	47°-30'	47°-30'	47°-30'

An air view of the world's largest iron ore docks. From left to right are Ore Docks One, Two, Three and Four. The photo was taken in the late 1940s. Ore Dock Two's approach has a C-1 080 steam engine pushing a shove of ore cars. Dock Three has another 080 and ore cars on it. Ore Dock Four is seen with a rail-mounted crane used to assist in breaking up ore and for helping make repairs on docks. ***Great Northern Railway, courtesy Burlington Northern Santa Fe Railway***

Pictured here is the wooden approach to Ore Dock One with the service maintenance building (shown on pages 71-75) in the upper right of the photo. ***Author's Collection***

Signal tower on Ore Dock One's approach controlled the traffic off and on all four ore docks, as the single track from the Allouez yard split into four approaches at the signal tower. This photo was taken prior to 1946, when the raised office service building (shown on pages 76-81) took control of that function. The steel approach to Ore Dock Two was 2,892 feet long, while the other three ore dock's approaches were made of wood and fed off of Two's approach. There was a .7 percent to 1 percent grade on the ore dock approach. ***Lee Thompson Collection***

Top view of the approach to the old wood Ore Dock Two. Tool and warming sheds are 80 feet above the water. The water barrel to right indicates concern from fire cinders being dropped from steam engines. *Great Northern Railway Historical Society*

Fire on wood approach at Ore Dock Two on January 31, 1922. One hundred firemen fought this huge blaze, which destroyed 600 feet of the wood approach and 14 pockets of Ore Dock Two, resulting in $60,000 damage. At the time of the fire, the Great Northern was to start dismantling the wooden approach and begin installing the steel approach. Ore Dock Two's fire did not discontinue Great Northern's service for the season. Great Northern went around the damaged approach with a temporary trestle for the 1922 ore season. *Unknown*

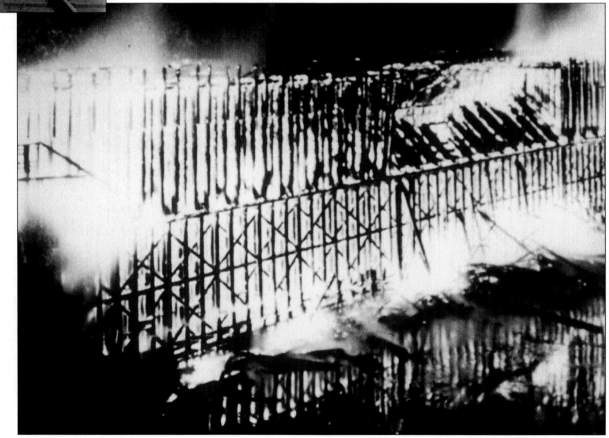

After the 1922 fire, a steel approach was built for Ore Dock Two. *Great Northern Public Relations, Burlington Northern Santa Fe Railway*

Another view of the steel approach to Ore Dock Two. The other three ore docks had wooden approaches. The office service building is seen to upper right. *Author's Collection*

19

Steel approach to Ore Dock Two continues onto the lake butting against cement columns. This view is seen from the walkway near the office service building. Note the walkway under the structure near lake level. *Author's Collection*

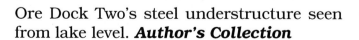

Ore Dock Two's steel understructure seen from lake level. *Author's Collection*

The approach for all four docks starts with one track and splits to two tracks. Shown is where the two tracks split to four tracks going onto Ore Dock Two. Gantry crane and Robbins car shaker (shown on pages 88-91) are above four sets of track at the beginning of Ore Dock Two. *Author's Collection*

There is a crossover on all four docks between the double tracks, allowing for easier maneuvering of ore cars. This photo was taken on Ore Dock Two. Track switch is at the left. *Author's Collection*

Drawings and dimensions of concrete and steel columns that hold Ore Dock One and Two's superstructure. Notice railing along fenders, snubbing posts and base construction. Ore Dock Four was of similar design. ***Great Northern Railway Blueprint, courtesy Burlington Northern Santa Fe Railway. Drawing by author***

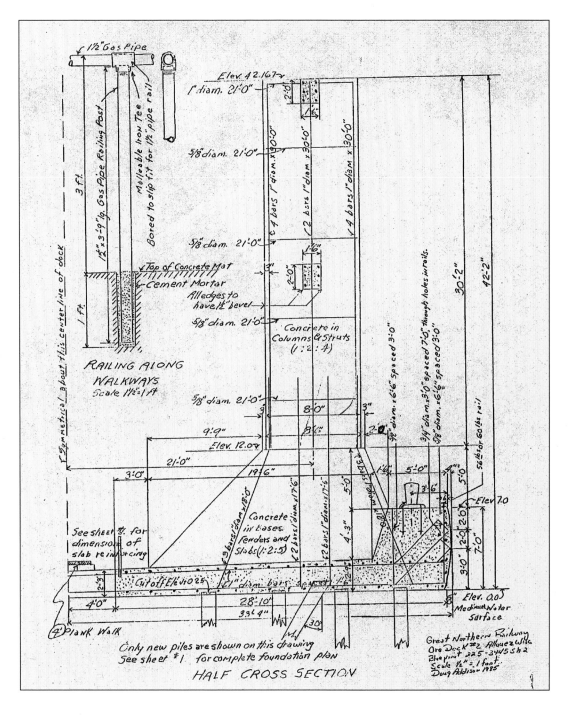

Long view of fender, snubbing posts, hand-rail, concrete columns, and life ring for the unlucky person who falls into the water. ***Author's Collection***

Under Ore Dock One: column base of concrete and steel. ***Author's Collection***

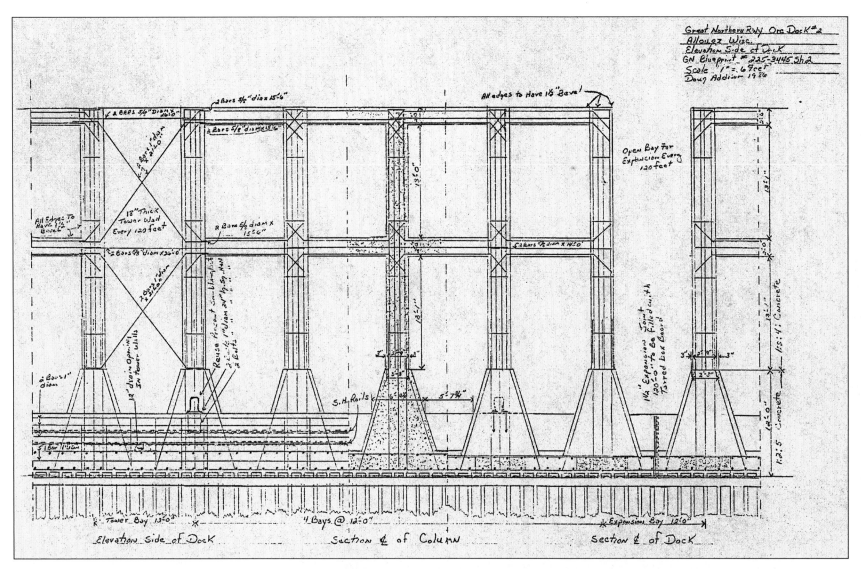

Side view cross-section of columns pertaining to Ore Docks One and Two. All columns are 12 feet center on center. Between the first two columns on the left is a concrete and steel tower that was 18 inches thick. There was one every 120 feet (10 columns). Between the last two columns on the right is an expansion with cut support between columns, one every 120 feet. Substructure has pilings underneath fender, handrails and snubbing posts. **Great Northern Railway Blueprint, courtesy Burlington Northern Santa Fe Railway. Drawing by author**

Here is a good view of the solid concrete and steel tower between columns on Ore Dock Two in the background. Spouts are in "up" position waiting for spring thaw. ***Author's Collection***

Under Ore Dock Two: the wood walkway to fender between column bases. Ore Dock One can be seen in background. There are towers of concrete and steel between some columns. The spouts are at 90 degrees. The superstructure steel supports are spaced 12 feet on center as are all the ore pockets. ***Author's Collection***

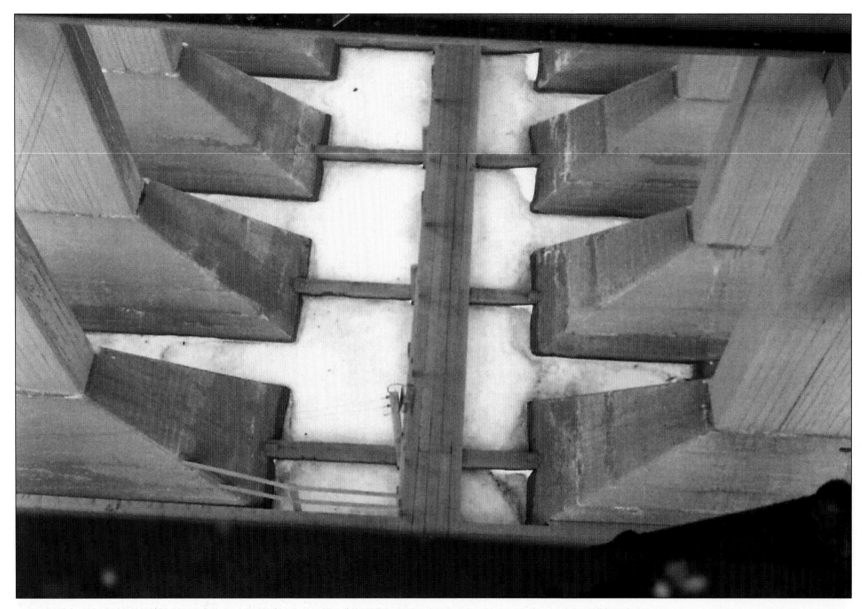

Under Ore Dock Two: walkway base of inner ore dock columns and wood walkway. Photo taken from catwalk on the end trestle (shown on page 30). *Author's Collection*

Under Ore Dock One looking up at electrical stand and electrical brace attached to column. *Author's Collection*

Walkway under Ore Dock Two: power house at the far end (shown on pages 82-84). Walkway steps to fender and water pipe attached to inside column base are seen to the left. *Author's Collection*

Under Ore Dock One electric wires run on each side for lights, spout machinery, and jackhammers. *Author's Collection*

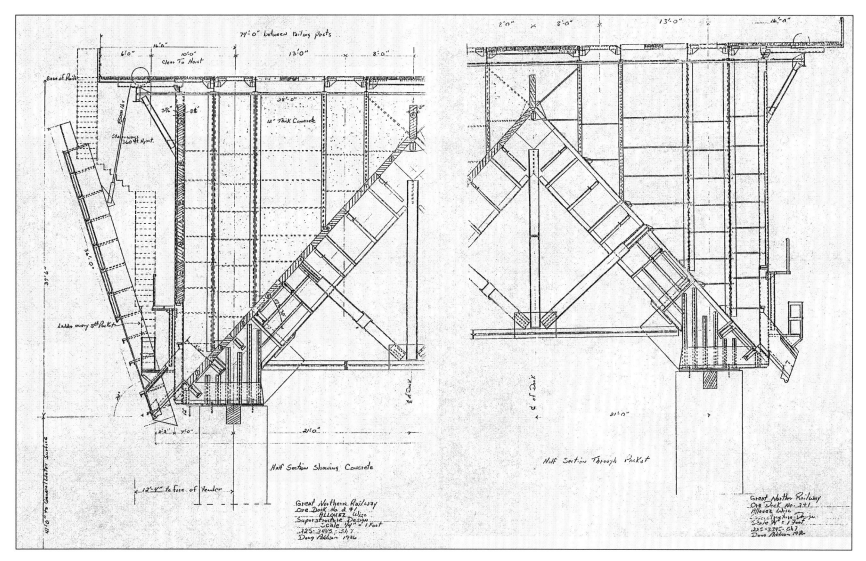

Cross-section of Ore Dock One and Two's upper superstructure. ***Great Northern Railway Blueprint, courtesy Burlington Northern Santa Fe Railway. Drawing by author***

Steel slanted I-beams support pockets and top of ore dock. Walkway (catwalk) at the end of the trestle is shown. *Author's Collection*

The steel supports hold the entire superstructure up by resting on top of the concrete columns. *Author's Collection*

Every 12 feet supporting fingers hold slanted I-beams that run the entire length of Ore Docks One and Two. *Author's Collection*

31

Looking up at the steel beams under superstructure that support I-beams spaced every 12 feet. These beams are responsible for holding up ore pockets and top deck. ***Author's Collection***

Looking down at the crossbeams and steel fingers that support I-beams. ***Author's Collection***

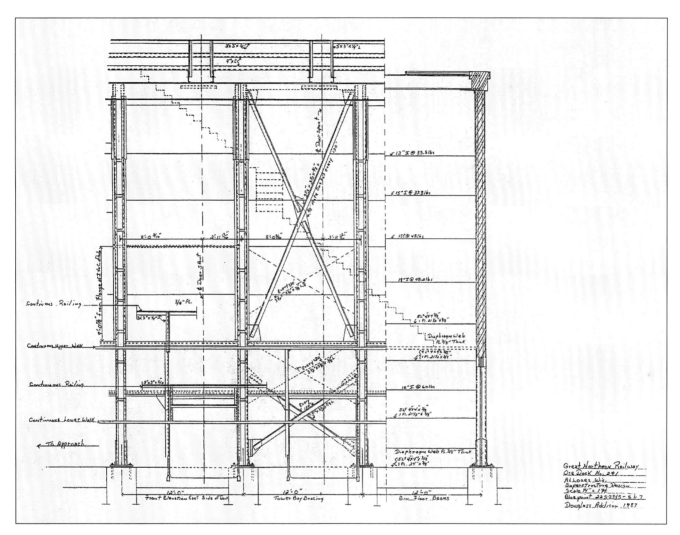

Side view of superstructure without spouts. Sides are concrete-with-steel beams spaced every 12 feet supporting ore pocket walls. Note the stairs from the top railing to the continuous catwalk. The catwalk runs just above the upper part of the ore pocket doors. Lower walkway provides easy access to each of the ore pocket doors. ***Great Northern Railway Blueprint, courtesy Burlington Northern Santa Fe Railway. Drawing by author***

East side of Ore Dock One. Two spouts on the left are down. Steel beams support pockets and outer wall. Both walkways and railings run the length of the dock. Seen under the first spout that is up on the left is an expansion space, also seen under the third spout from the right. The middle support columns have concrete filled in for added support between them and are called "towers." ***Author's Collection***

Stairway from top of Ore Dock One down to continuous catwalk. Cables are for ore pocket doors. Spouts are down and kept in place by Ribbon steel harness and Ribbon cable. ***Author's Collection***

Ladder down to catwalk that has access to ore pocket doors. On this cold and quiet day on the docks, the only sounds heard were the occasional pinging of small concrete pieces hitting the steel structure as they broke and fell due to contraction in the minus 20-degree weather. A few concrete chips are seen here along the catwalk. **Author's Collection**

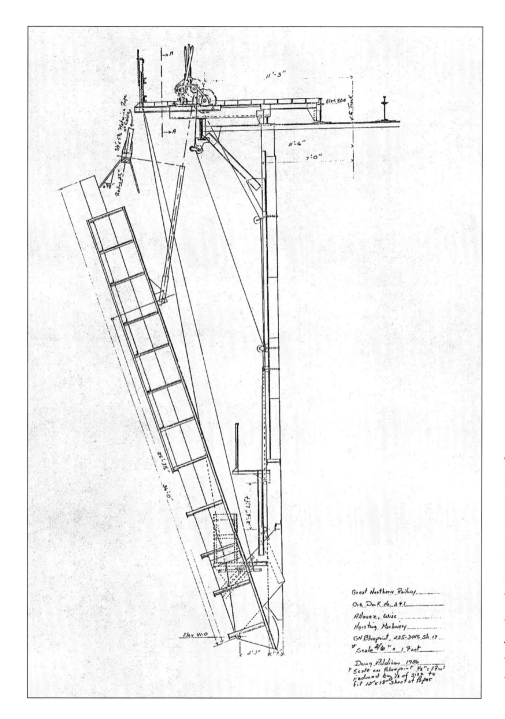

Great Northern Railway
Ore Dock No. 2 & 1
Allouez, Wisc
Hoisting Machinery
GN Blueprint 225-3445 Sh. 17
Scale 3/6" = 1 Foot
Dwg Addition 1986
Scale on Blueprint 1/2" = 1 Foot
Reduced by 1/3 of size to
fit 12"x18" Sheet of Paper

This drawing shows the ore dock overhang with hoisting machinery. Cables and rollers are attached to the ore pocket door spout connection and hoist. This was similar for all four docks and everything was 12 feet on center. ***Great Northern Railway Blueprint, courtesy Burlington Northern Santa Fe Railway. Drawing by author***

Top of ore dock overhangs with angle iron support and guide roller cable leading to spout machinery. ***Author's Collection***

Rollers for pocket door cables are spaced every 12 feet on center. Overhang is wood supported by angled iron supports. ***Author's Collection***

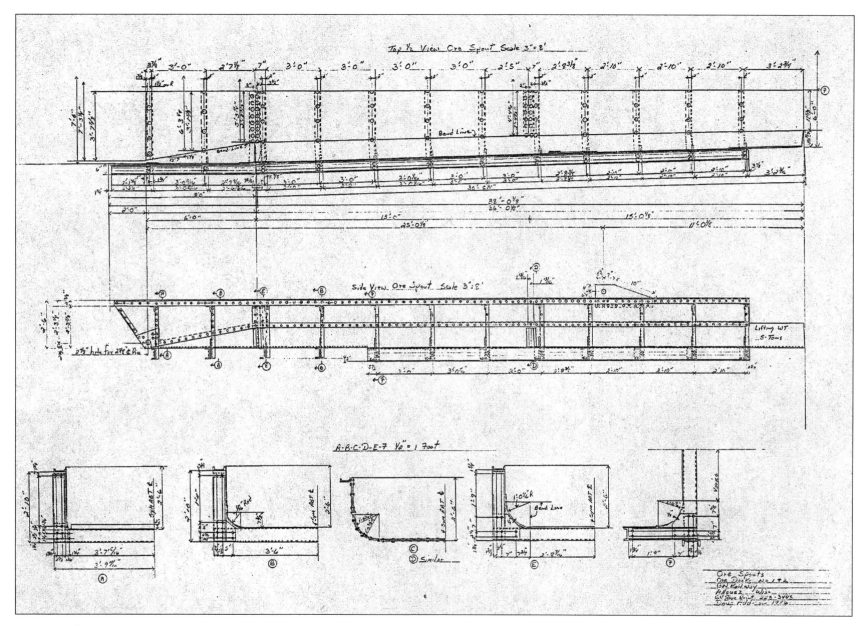

Top half view and side view of spouts for Ore Docks One and Two. The spouts are 36 feet long and six feet wide at the end of the spout where ore pours into the ore boat hold. *Great Northern Railway Blueprint, courtesy Burlington Northern Santa Fe Railway. Drawing by author*

Spouts are at 90-degree angles. For unknown reasons these spouts were lowered on Ore Dock One in the winter and raised during the shipping season. **Author's Collection**

Below: Ore spouts are in up position and are held in place by harness and cables attached to hoisting machinery on overhang. ***Author's Collection***

Above: Spouts are in the "up" position. Spouts are ***supposed*** to be even numbered on one side of the ore dock and odd numbered on the other side. These spouts are not in sequence. As spouts were damaged they were replaced with others previously taken out of service. ***Author's Collection***

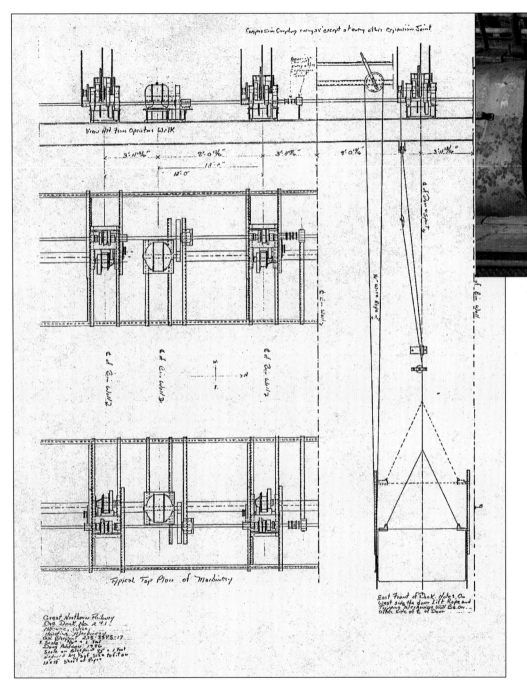

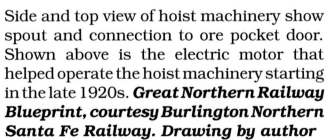

Side and top view of hoist machinery show spout and connection to ore pocket door. Shown above is the electric motor that helped operate the hoist machinery starting in the late 1920s. *Great Northern Railway Blueprint, courtesy Burlington Northern Santa Fe Railway. Drawing by author*

Shown is the drive shaft that connects to electric motor that operates hoist machinery. *Author's Collection*

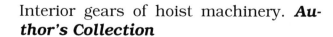

Interior gears of hoist machinery. *Author's Collection*

Hoist machinery lined up on one side of the dock. *Lee Thompson Collection*

Hoist machinery is spaced every 12 feet on center of ore pockets and ore pocket doors. *Author's Collection*

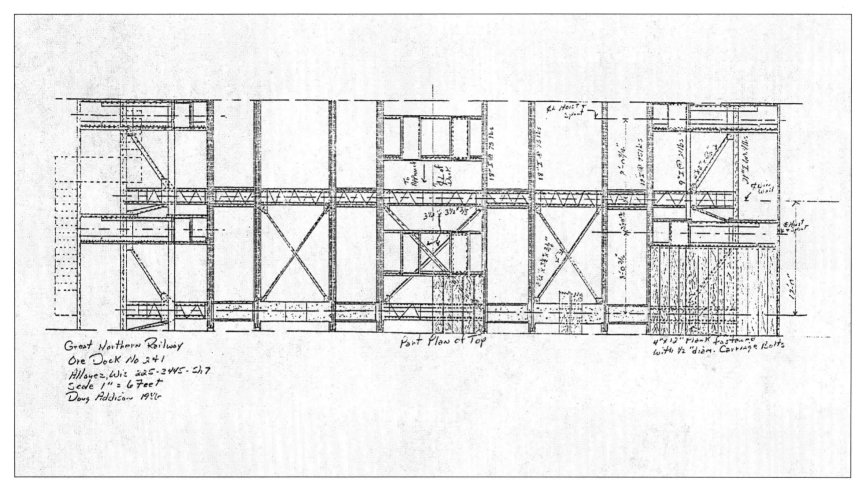

Top view Ore Docks One and Two have four sets of tracks; two sets of tracks are on each side of docks. The pockets are slanted 47 degrees for 30 feet and positioned 12 feet center on center. There is a walkway down the middle of the dock from the beginning to the end of the trestle. On the outer edges are hoist machines 12 feet center on center for each pocket and ore pocket door. There were 374 ore pockets on Ore Dock One, 350 on Ore Dock Two, 236 on Ore Dock Three and 202 on Ore Dock Four; for a total of 1,352 pockets that could hold 441,900 tons of iron ore. ***Great Northern Railway Blueprint, courtesy Burlington Northern Santa Fe Railway. Drawing by author***

A steel walkway is in the center of Ore Docks One and Two. *Author's Collection*

Steel posts support the handrail between pockets; there are two sets of tracks on each side of the handrail. *Author's Collection*

Inside rails over ore pockets. Pocket walls are located 12 feet on center and are made of concrete and steel with bars going front-to-back and side-to-side to break up ore being dumped into pockets from ore cars. These steel crossbars were only installed on Ore Dock One. *Author's Collection*

Seen on Ore Dock Two, a concrete walkway exists between tracks and rail construction. Pocket walls have steel supports on sides. *Author's Collection*

Ore pockets are all pre-stressed concrete walls and the slanted wall in the middle moved ore downward by gravity. All the ore docks had pockets that were slanted at 47 degrees for 30 feet. ***Author's Collection***

Pocket door on base of outside wall emptied ore out of pockets and onto spouts. ***Author's Collection***

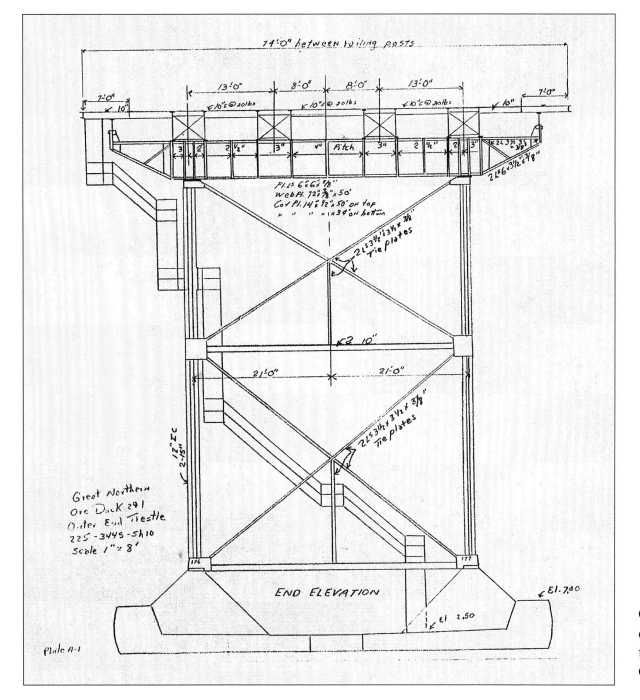

Cross-section view of the front of the end trestle on Ore Docks One and Two.

49

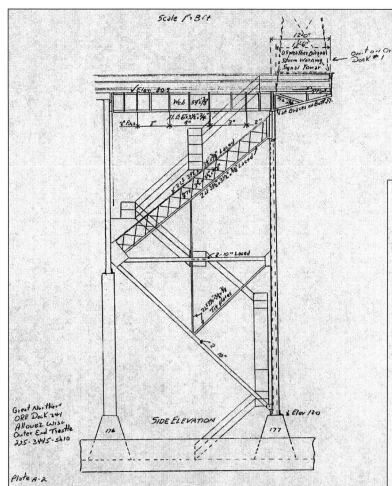

Side view of end trestle of Ore Dock Two (Ore Dock One and Four were identical but had no weather station; Ore Dock Three, of wood construction, also had no weather station). *Great Northern Railway Blueprint, courtesy Burlington Northern Santa Fe Railway. Drawing by author*

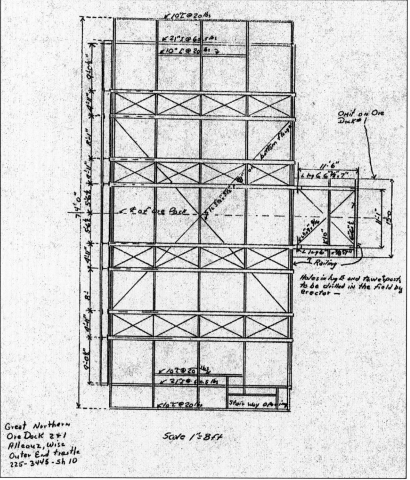

Top grid work for end trestle of Ore Dock Two (Ore Dock One had weather station). *Great Northern Railway Blueprint, courtesy Burlington Northern Santa Fe Railway. Drawing by author*

End trestle of Ore Dock One. ***Author's Collection***

End Trestle of Ore Dock Two. ***Author's Collection***

Four sets of tracks come to an end at the end trestle of Ore Dock Two. *Author's Collection*

The tracks curl up backed by large timbers to keep ore cars from going off the end of the trestle. White shack is where U.S. Weather Bureau Tower used to be on Ore Dock Two. *Author's Collection*

Sheet-steel toilet and wood maintenance shed on west side of Ore Dock Two's end trestle. *Author's Collection*

Ore Dock Two's end trestle stairway to mattress (base). *Author's Collection*

Stairway midway down from top of end trestle. **Author's Collection**

Stairway at bottom of end trestle. **Author's Collection**

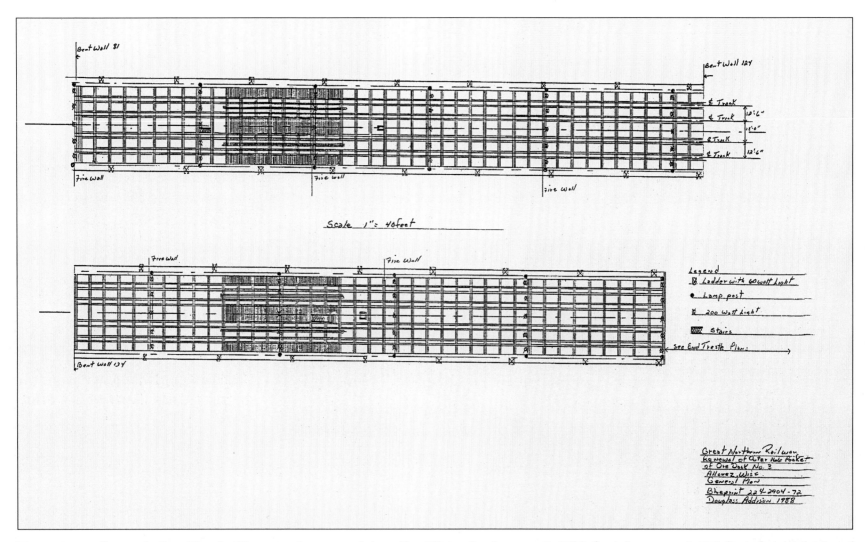

Top view of wood Ore Dock Three minus end trestle. This dock was 1,956 feet long and 77 feet from deck of dock to water. *Great Northern Railway Blueprint, courtesy of Burlington Northern Santa Fe Railway. Drawing by author*

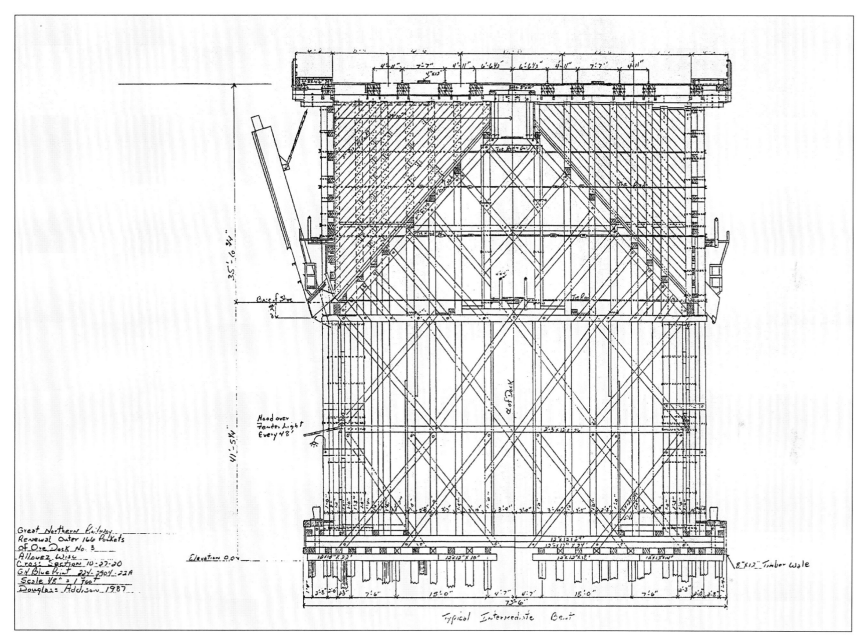

Cross-section of wood Ore Dock Three. This dock had 326 pockets and spouts. *Great Northern Railway Blueprint, courtesy of Burlington Northern Santa Fe Railway. Drawing by author*

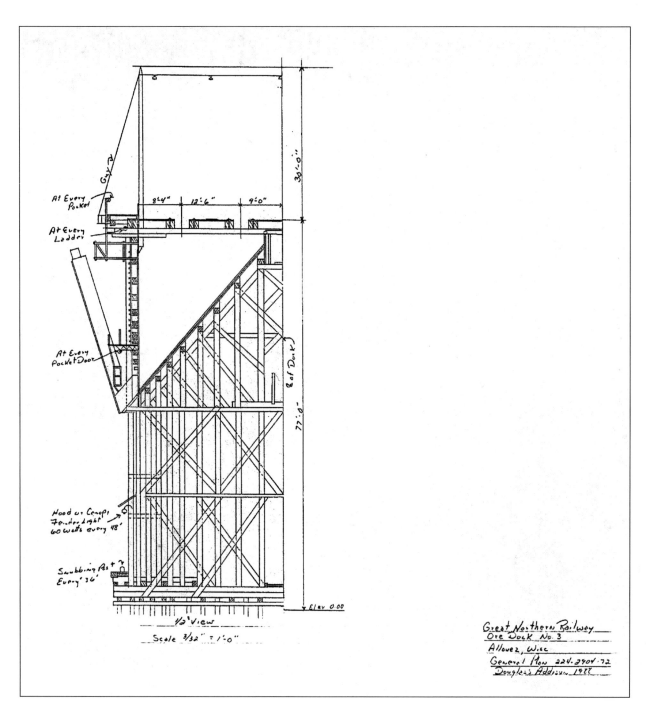

At Every Pocket

At Every Ladder

At Every Pocket Door

Guy P

8'4" 12'6" 9'0"

30'-0"

Dk of Dock

77'-0"

Hood w Canopy Fender Light 60 watts every 48'

Snubbing Post Every 36'

Elev 0.00

1/2 View
Scale 3/32" = 1'-0"

Great Northern Railway
Ore Dock No. 3
Allouez, Wisc
General Plan 224.2904.72
Douglas's Addition 1922

Half view of wood Ore Dock Three. Again, it was 77 feet from water line to top of rails on deck top. It stretched 30 feet higher for the light poles and five lights were strung across the deck side-to-side. *Great Northern Railway Blueprint, courtesy of Burlington Northern Santa Fe Railway. Drawing by author*

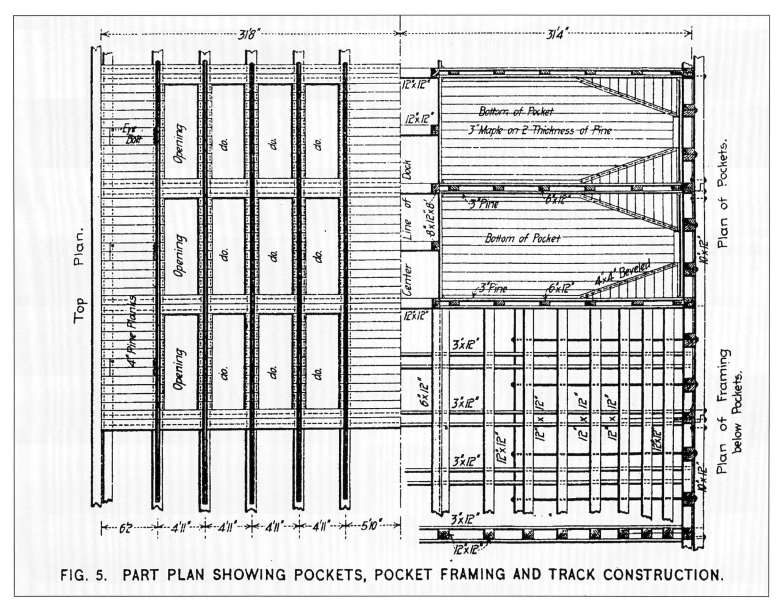

FIG. 5. PART PLAN SHOWING POCKETS, POCKET FRAMING AND TRACK CONSTRUCTION.

Top deck Ore Dock Three. Port plan showing pockets, pocket framing and track construction. ***Great Northern Railway Historical Society***

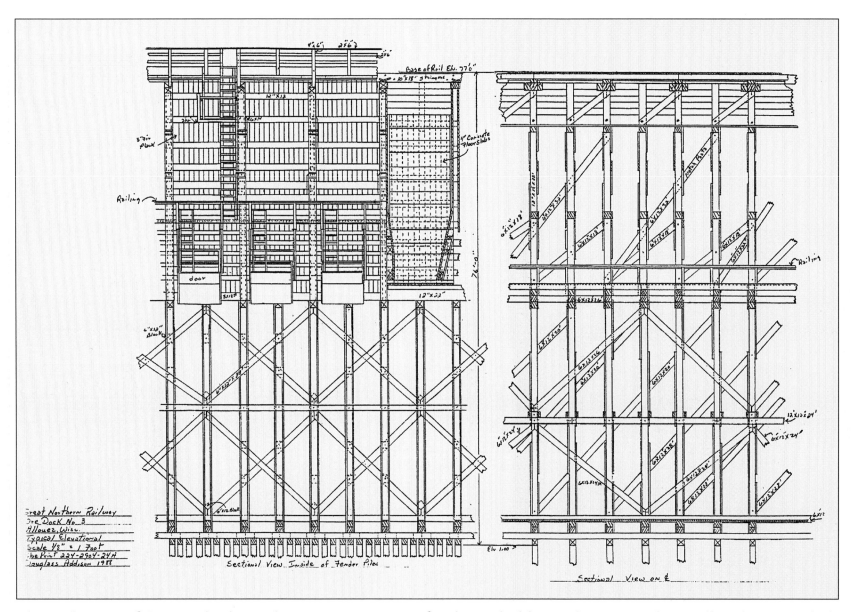

Sectional View Inside of Fender Piles

Sectional View on ℄

This side view of Ore Dock Three shows construction of railings, ladder and outer pocket walls. There is a ladder to upper walkway and ore pocket doors. The inside view of the pocket is tapered at the ore pocket door. The substructure is wooden to fender and pilings are below fender. ***Great Northern Railway Blueprint, courtesy of Burlington Northern Santa Fe Railway. Drawing by author***

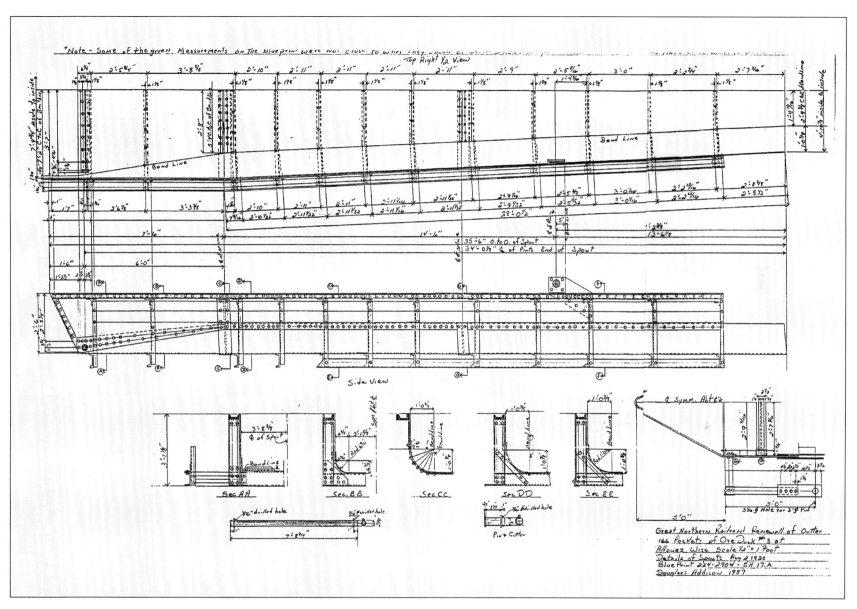

Side and half top view of ore spout on Ore Dock Three. At 34 feet long it was the smallest of spouts on the four docks. There were 326 spouts, pockets and pocket doors. Spout machinery and two sets of tracks were located on each side of the dock. *Great Northern Railway Blueprint, courtesy of Burlington Northern Santa Fe Railway. Drawing by author*

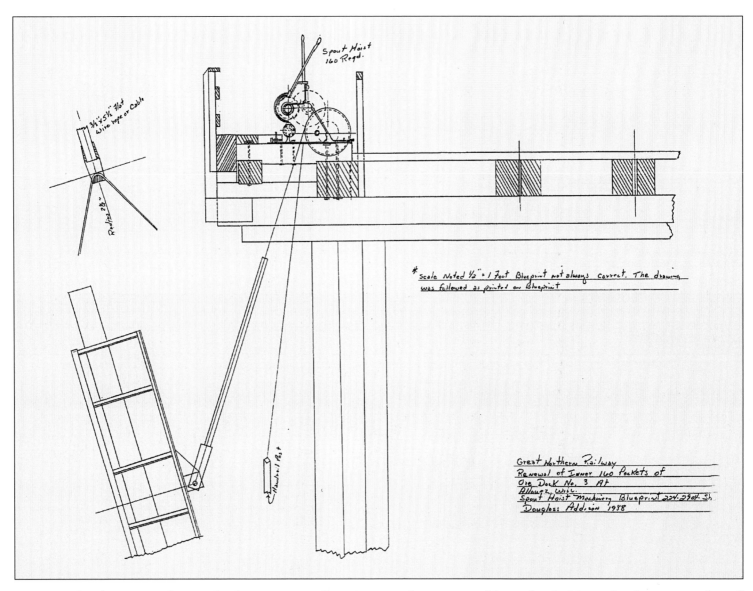

Ore Dock Three overhang deck supported spout machinery and handrail. Note the harness detail and cable hookup to spout from spout machine. When ore went down spouts to ore boat's hold, the spout machinery operators could raise and lower the spout to control the level of ore in the hold. ***Great Northern Railway Blueprint, courtesy of Burlington Northern Santa Fe Railway. Drawing by author***

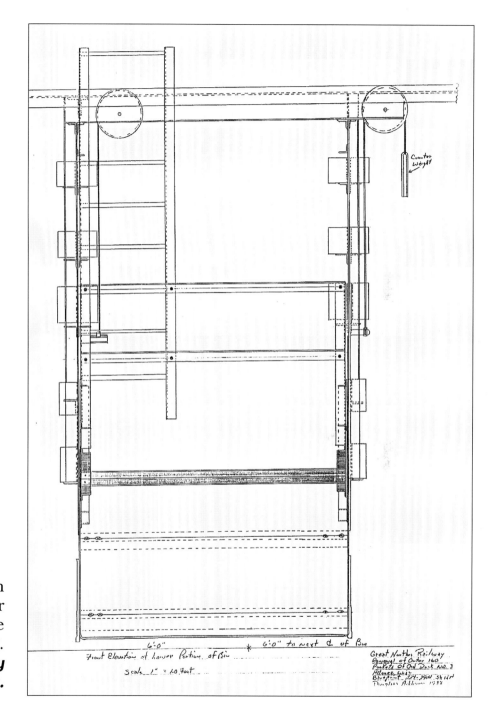

Ore Dock Three's ladder from mid-walkway down to ore pocket door. Also notice ore pocket door with cable and counterweight. The pockets were maple and had a pre-stressed concrete covering. **_Great Northern Railway Blueprint, courtesy of Burlington Northern Santa Fe Railway._**

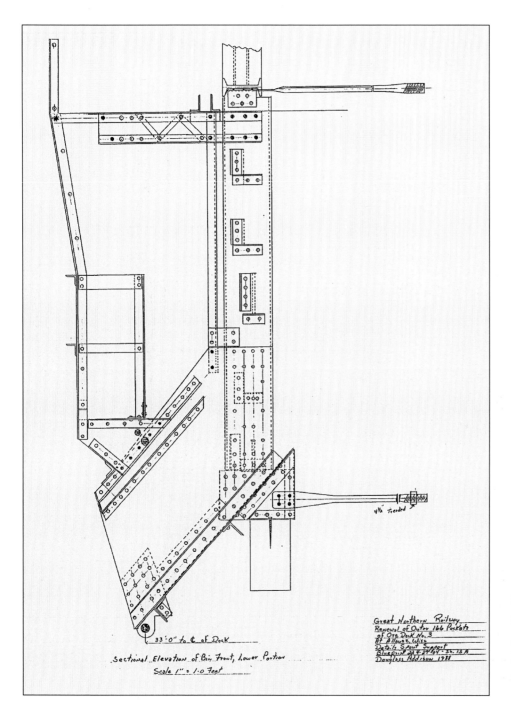

Great Northern Railway
Renewal of Outer 166 Pockets
of Ore Dock No. 3
At Allouez, Wis.
Details Spout Support
Blueprint #224-24/63 - Sh. 13 A
Douglass Addison 1988

33'-0" to ℄ of Dock

4½" Treaded

Sectional Elevation of Bin Front, Lower Portion
Scale 1" = 1.0 Feet

Ore Dock Three's spout support where spout attaches to mule at lower left. ***Great Northern Railway Blueprint, courtesy of Burlington Northern Santa Fe Railway. Drawing by author***

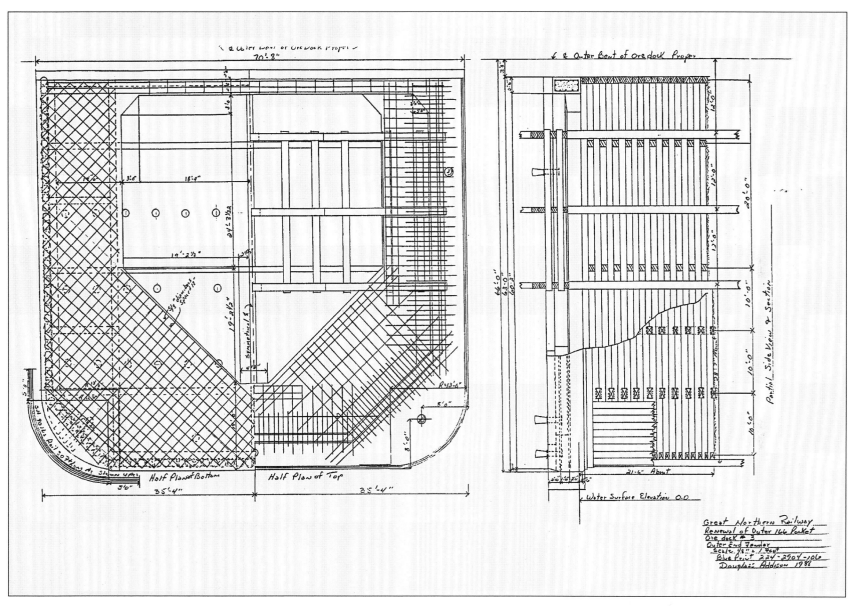

Drawing of Ore Dock Three's end trestle—half view plan of top deck and half view plan of bottom construction (left); side view of end trestle fender and below water substructure (right). ***Great Northern Railway Blueprint, courtesy of Burlington Northern Santa Fe Railway. Drawing by author***

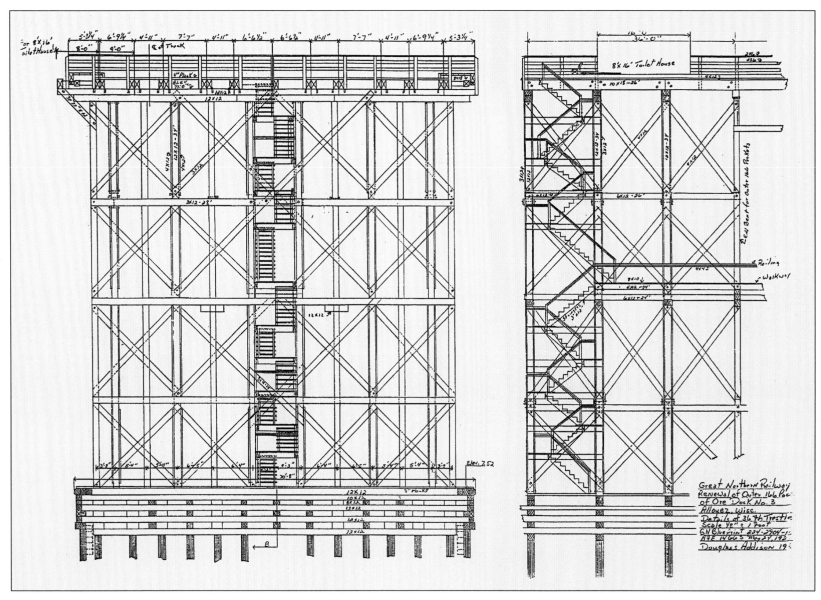

View of end trestle of Ore Dock Three from dock top and railing to the fender and substructure. The ladder is to the front of the end trestle. Side view of end trestle shows toilet house on top deck and stairway from deck top to mattress. Halfway up the end trestle, a walkway connects the continuous catwalk to the stairway. Great *Northern Railway Blueprint, courtesy of Burlington Northern Santa Fe Railway. Drawing by author*

Both photographs show what is left of Ore Dock Three. The cut-down end trestle (above) and cut-off bases of superstructure supporting timbers (left) are shown here above ice and snow. ***Author's Collection***

The only original steel and concrete ore dock. Ore Dock Four went into service in 1911. Ore Docks One and Two were originally wood and were converted to steel and concrete by 1928. The dock would not accommodate a Robbins car shaker. A level car shaker was used on Ore Docks Three and Four. Ore Dock Four was 1,812 feet long and the deck was 75 feet above the water. Spouts were 34 feet 6 inches long and the angle of the pockets was 47 degrees for 30 feet. There were 302 pockets and spouts.

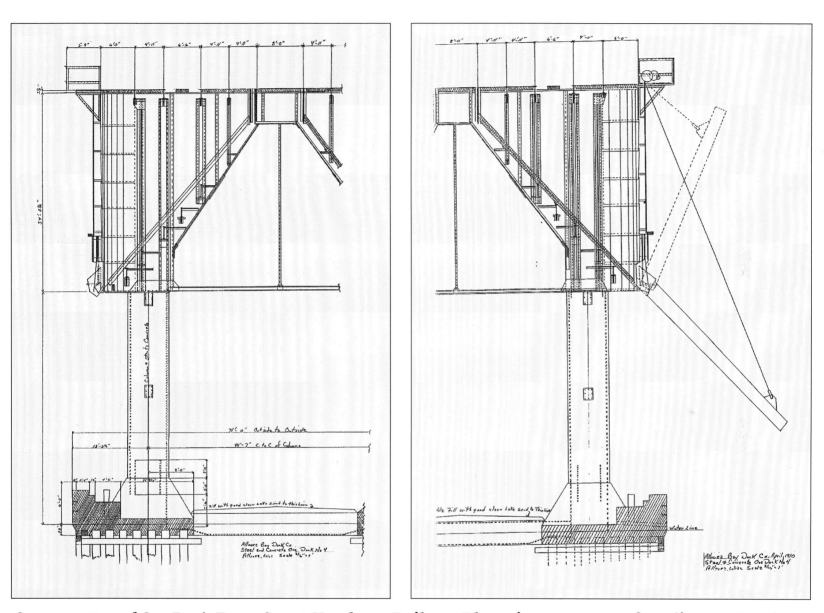

Cross-section of Ore Dock Four. *Great Northern Railway Blueprint, courtesy of Burlington Northern Santa Fe Railway. Drawing by author*

Ore flowing down spout from open pocket door by operator on Ore Dock Four. *Great Northern Railway Photo, courtesy of Burlington Northern Santa Fe Railway*

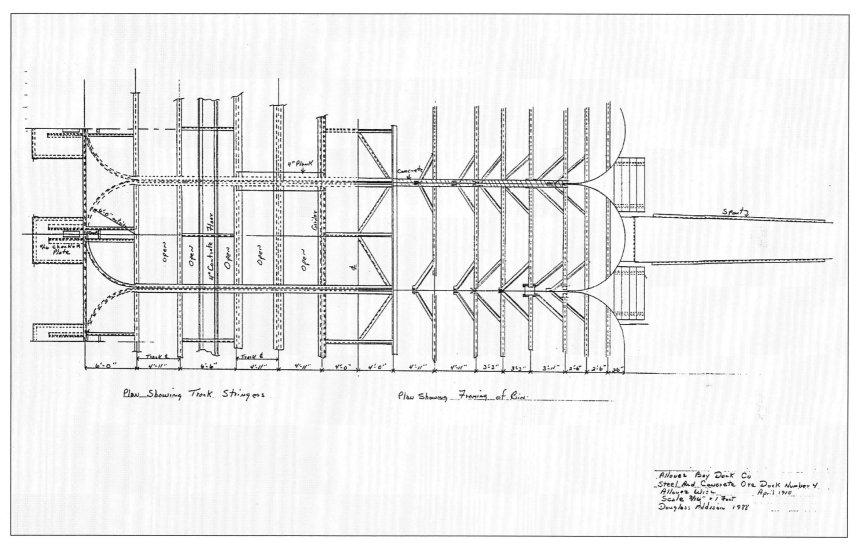

Top dock view of Ore Dock Four showing crank stringers and framing of box. Note the barrel-construction of the ore pockets. Like the other three docks, Ore Dock Four had four sets of tracks; two on each side. *Great Northern Railway Blueprint, courtesy of Burlington Northern Santa Fe Railway. Drawing by author*

East side of Ore Dock Four without spouts. Ore Dock Two is seen in background and most of the spouts are off. When this photo was taken in the late 1980s, Ore Docks Two and Four were starting to be taken apart. *Author's Collection*

Side view of Ore Dock Four to stairway on end, and view of dock minus spouts. The superstructure is different from the other three ore docks. The top deck railing, ladder, steel outer barrel pocket walls, ore pocket doors, counterweights and spouts are all different. The substructure is the same as Ore Docks One and Two. *Lee Thompson Collection*

Maintenance service building on legs of steel elevated to walkway level seen from the southeast side. Ice tower from melting snow extends from northeast corner of building to ground. ***Author's Collection***

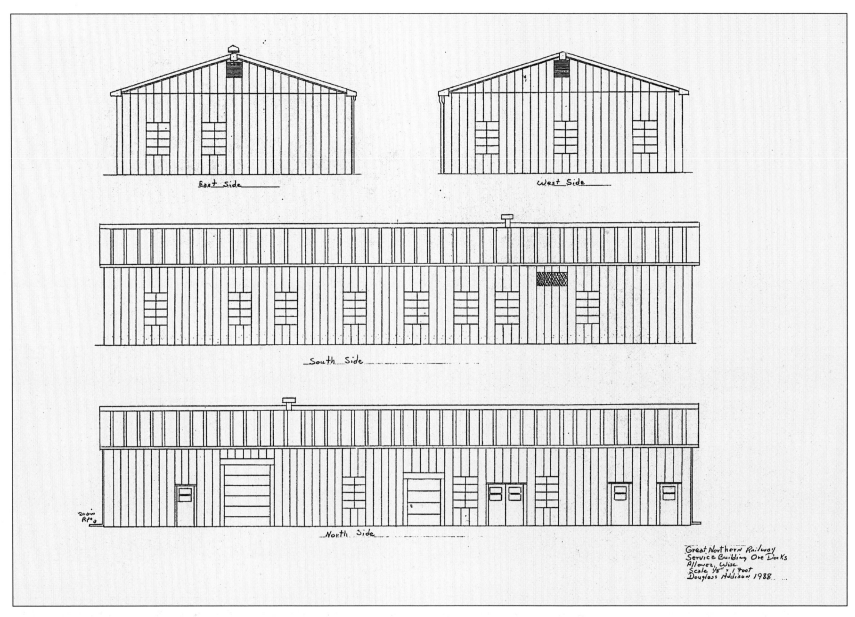

East Side

West Side

South Side

North Side

Great Northern Railway
Service Building Ore Docks
Allouez, Wisc.
Scale 1/8" = 1 foot
Douglass Addison 1988

Drain Pipe

Drawing, by author, of Great Northern maintenance service building between Ore Docks One and Two. Measurements were taken by Lee Thompson and author.

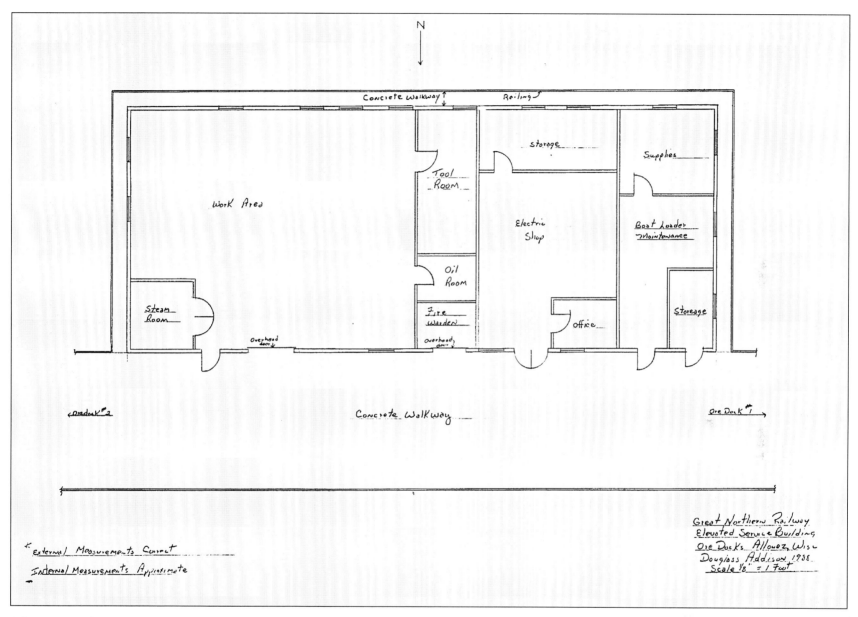

N

Concrete Walkway ↑   Railing

Tool Room

Work Area

storage

Supplies

Electric Shop

Boat Loader Maintenance

Oil Room

Steam Room

Fire Warden

Office

Storage

Overhead door ↓

overhead door

Ore Dock #2

Concrete Walkway

Ore Dock #1

External Measurements Correct

Internal Measurements Approximate

Great Northern Railway
Elevated Service Building
Ore Docks Allouez, Wisc.
Douglass Addison 1988
Scale ⅛" = 1 Foot

Shown is the Great Northern maintenance service building's concrete walkway in front, and concrete and walkway with railing around other three sides of service building. The interior and outer measurements were taken by Lee Thompson and author.

The maintenance service building seen from the southwest corner shows the concrete walkway around the building for easy access. *Author's Collection*

Dock foreman is riding gas-powered cart on Ore Dock One. Maintenance service building faces north toward the loading section of the ore docks. At one time 400 workers were needed to operate the four docks. In the mid-1980s only a foreman and two ore-loading employees were needed to operate the docks. *Author's Collection*

Front of maintenance service building in the early 1980s. A gas pump is next to railing for gas-powered cart. **Author's Collection**

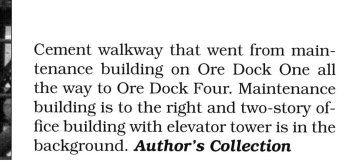

Cement walkway that went from maintenance building on Ore Dock One all the way to Ore Dock Four. Maintenance building is to the right and two-story office building with elevator tower is in the background. **Author's Collection**

The office service building was completed by November 1945 and in use before the end of the shipping season in 1945. This two-story office service building's top floor is 84 feet above the water. It is supported by steel girders on a concrete base, which in return is supported by wooden pilings set 60 feet into the ground. Situated between Ore Docks Two and Three, the building is the nerve center for all operations on the four docks including controlling train movement to and from the Iron Range and the routing of incoming boats to the different docks. Great Northern employees had a complete view of all four ore docks. A ten-passenger automatic elevator, the shaft and tower at the end of the building was installed to save the office personnel and boat captains (who had papers to sign) the tedious task of walking from ground level up the stairway next to the elevator to office level. The two-story building is 132 feet long and 30 feet wide. The small brick building on the water level, under the office building, is a septic system that runs pipes from the office building down to the septic tank building.
Great Northern Public Relations, Great Northern Railway Historical Society

Closer view of north side (lake) of two-story office service building. ***Author's Collection***

The two-story brick and cement general office building housed the locker room, toilet washroom and lunch room on the bottom level. The elevator tower extends above northwest side and elevator shaft runs from the second floor to lake level and steel stairway is next to elevator housing. Brick septic tank building is seen at lake level. ***Author's Collection***

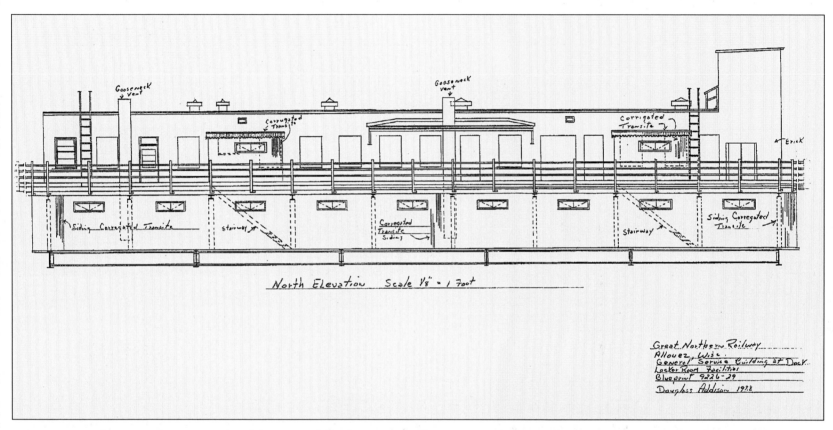

Blueprint of general service building and locker room facilities. The top floor is office spaces and the bottom level is ore dock workers' facilities. The elevator is located at the northeast corner (right side). *Great Northern Railway, courtesy Burlington Northern Santa Fe Railway. Drawing by author*

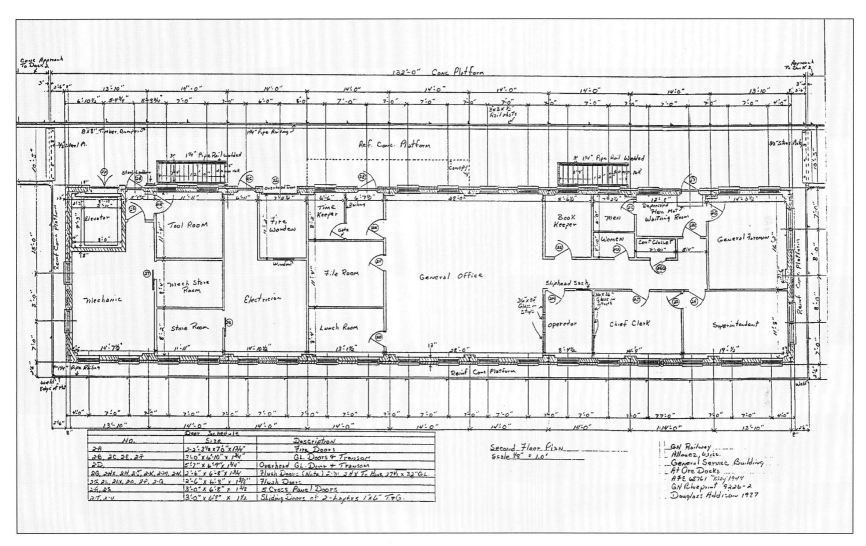

Great Northern blueprints of general office spaces on the top level. ***Great Northern Railway, courtesy Burlington Northern Santa Fe Railway***

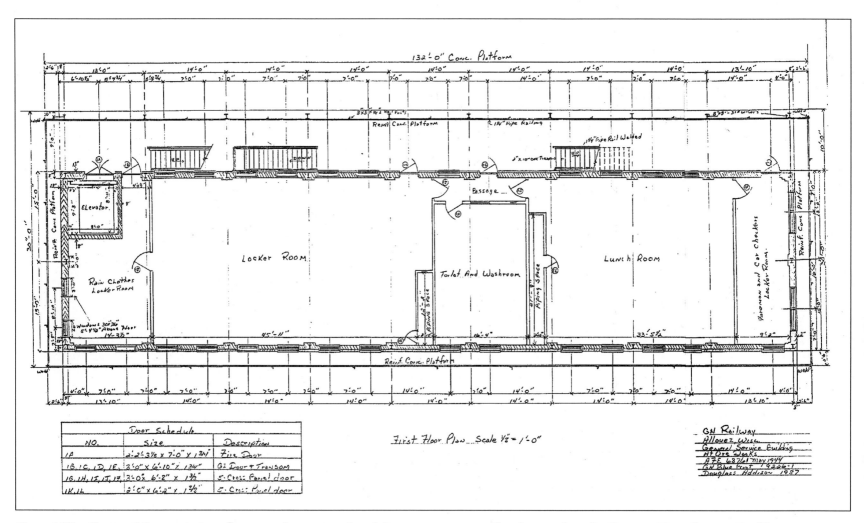

Great Northern blueprints of general service building spaces on the lower level. *Great Northern Railway, courtesy Burlington Northern Santa Fe Railway. Drawing by author*

Septic tank and house under general service building. Note crooked sewer pipe to septic house from general service building. ***Author's Collection***

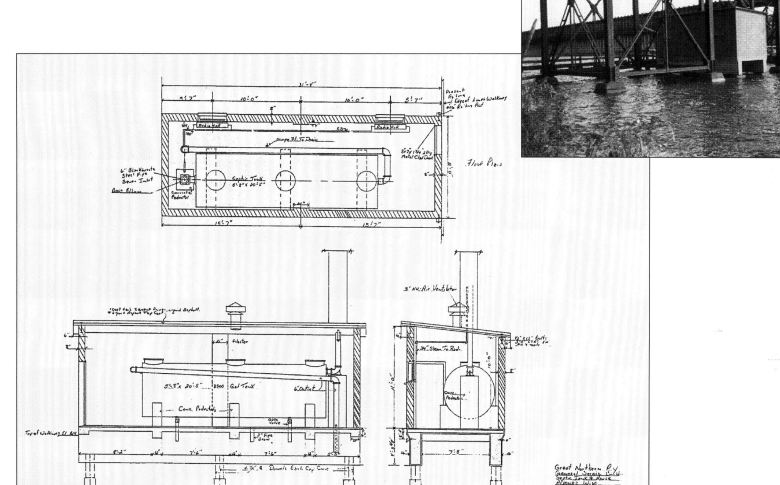

Great Northern blueprint of septic tanks and house under general service building. ***Great Northern Railway, courtesy Burlington Northern Santa Fe Railway. Drawing by author***

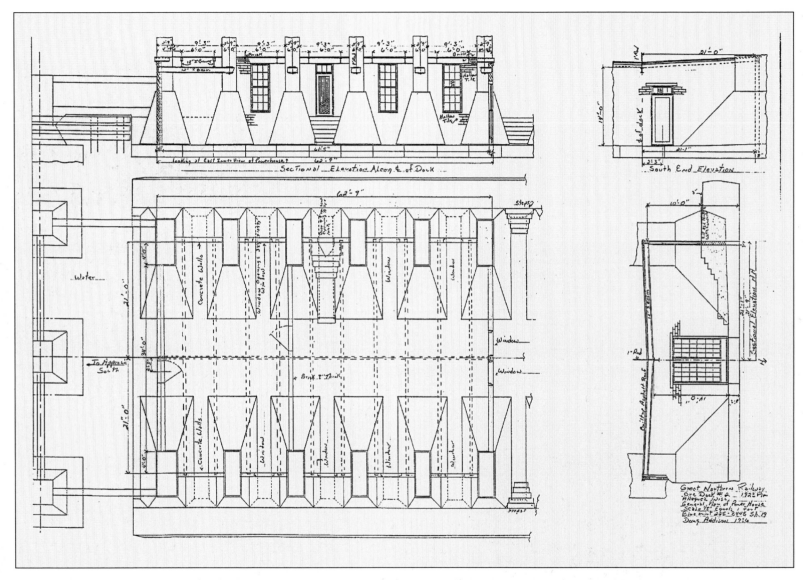

Great Northern blueprint of top, side and end view of power house under front of Ore Dock Two. Completed by 1923, the power house had electricity from another source and distributed the power to all dock's electrical systems. *Great Northern Railway, courtesy Burlington Northern Santa Fe Railway. Drawing by author*

Power house is seen in the lower right under Ore Dock Two between tower and expansion space of columns. Note steel stairway from base to upper deck on right. ***Author's Collection***

Rooftop of power house; left side seen from south.
Author's Collection

Rooftop of power house; right side seen from south.
Author's Collection

Doorway from shore side to power house. *Author's Collection*

Lakeside (north side) of power house under Ore Dock Two. *Author's Collection*

Pump house under Ore Dock Two pumped 800,000 gallons per day to help keep ore dust out of air. *Author's Collection*

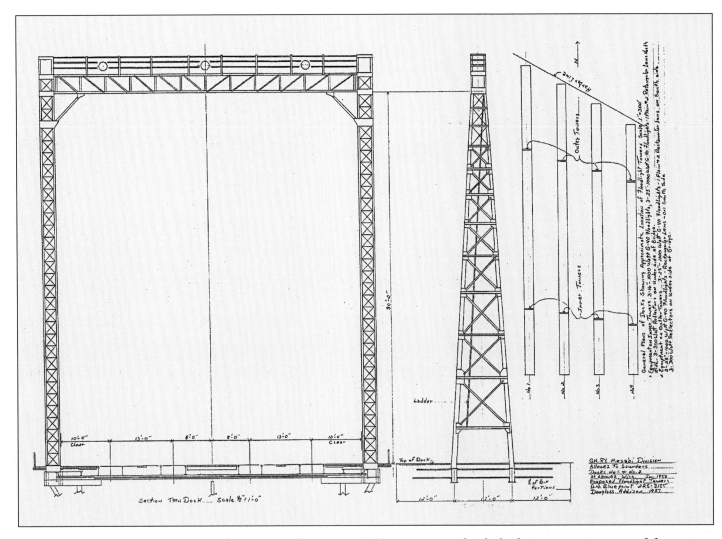

Great Northern blueprint of proposed, post-1952 new ore dock lighting system and location of the two lighting towers on Ore Docks One, Two, Three and Four. The lighting system was installed on Ore Docks One and Two, but the author is unsure if they were installed on Ore Docks Three and Four. *Great Northern Railway, courtesy Burlington Northern Santa Fe Railway. Drawing by author*

Lighting tower on Ore Dock Two is 80 feet high above deck. In 1952 and 1953 lights were installed on Ore Docks One and Two.
Author's Collection

Side view of light tower ladder to top on inside providing access to the top and walkway to replace burned out lights.
Author's Collection

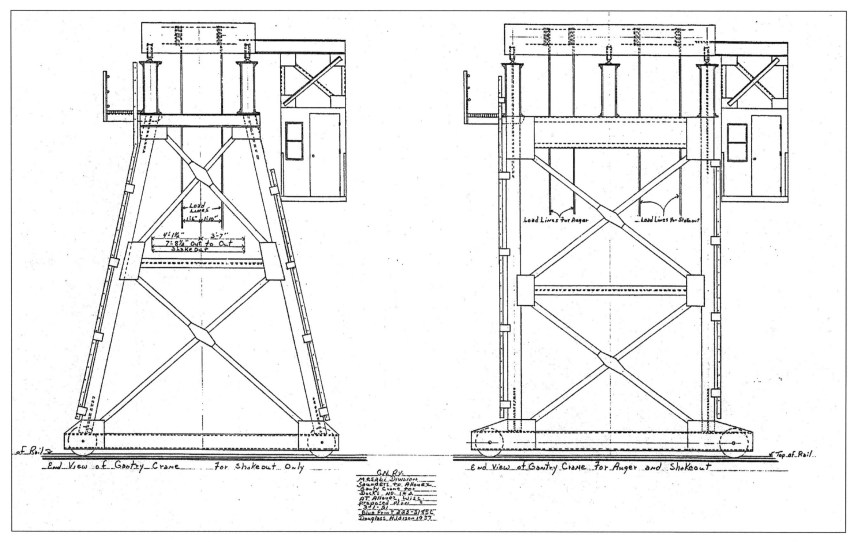

Great Northern blueprint of proposed Gantry crane. End view of Gantry crane for shakeout only and end view for Gantry crane for auger and shake out. Installed on Ore Dock One, 1952-1953: installed on Ore Dock Two, 1953-1954. Ore Docks Three and Four did not utilize the Gantry crane. ***Great Northern Railway Blueprint, courtesy Burlington Northern Santa Fe Railway. Drawing by author***

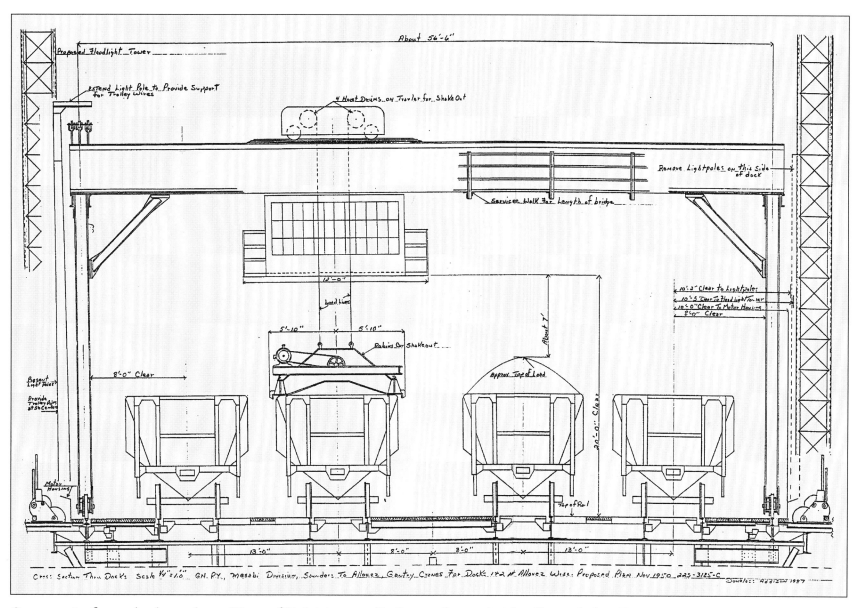

About 56'-6"

Proposed Floodlight Tower

Extend Light Pole to Provide Support for Trolley Wires

4 Hoist Drums on Trawler for Shake Out

Remove Lightpoles on this side of dock

Service Walk for Length of bridge

12'-0"

Load Lines

10'-2" Clear to Lightpole
10'-5" Clear to Flood light Tower
10'-0" Clear To Motor Housing
8'-0" Clear

5'-10"   5'-10"

Robins Car Shakeout

About 7'

8'-0" Clear

Approx Top of Load

Present Light Point

Provide Trolley Arm at 56 Center

20'-0" Clear

Motor Housing

Top of Rail

13'-0"   8'-0"   8'-0"   13'-0"

Cross Section Thru Docks Scale 1/4"=1'-0" GN. RY., Mesabi Division, Sounders To Allouez, Gantry Cranes For Docks 1 & 2 at Allouez Wisc Proposed Plan Nov 1950 225-3125-C   Douglas Addison 1987

Crosscut of ore dock and position of light tower. Rails on the outside allowed the Gantry crane to run the length of Ore Docks One and Two. Gantry crane holds Robbins car shaker that rode on rails and provided access to ore cars on all four sets of tracks over ore pockets. *Great Northern Railway, courtesy Burlington Northern Santa Fe Railway. Drawing by author*

Clam attached to the ore car. Robbins car shaker could shake ore loose at the rate of 40 ore cars per hour or 600 to 800 cars per 24-hour period. Every pocket on Ore Dock's One and Two were accessible. *Great Northern Railway, courtesy Burlington Northern Santa Fe Railway*

Clam from Robbins car shaker on ore car. Crane on tracks uses attachment ball to break up ore in ore cars. Shown here after 1953; note new lighting system in background. *Author's Collection*

Gantry crane wheels on front and back rest on rails that run the length of Ore Docks One and Two. **Author's Collection**

Robbins car shaker clam that raised, held and shook iron ore out of the bottom of ore cars and down into ore pockets. This procedure was also effective on frozen ore in late spring and late fall. **Author's Collection**

Pedestal Crane

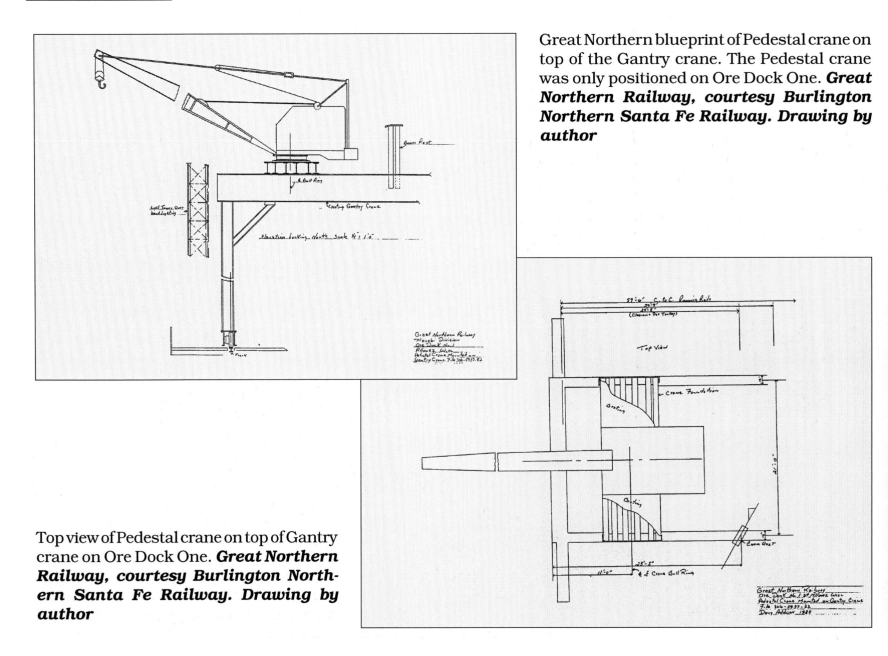

Great Northern blueprint of Pedestal crane on top of the Gantry crane. The Pedestal crane was only positioned on Ore Dock One. *Great Northern Railway, courtesy Burlington Northern Santa Fe Railway. Drawing by author*

Top view of Pedestal crane on top of Gantry crane on Ore Dock One. *Great Northern Railway, courtesy Burlington Northern Santa Fe Railway. Drawing by author*

Pedestal crane and Robbins car shaker on top of Gantry crane on Ore Dock One. Covered taconite belt (explained on pages 94-97) on left leading to pocket loading equipment. Pedestal crane rode side to side on Gantry crane. Gantry crane and Robbins car shaker rode on rails the length of Ore Docks One and Two. *Author's Collection*

Conversion to Taconite

From the 1890s, the Mesabi Iron Range in Minnesota supplied iron ore to the Great Northern Railway. The best grade ore (class one) had 50 percent iron or more, and processing was not needed. The class-two low-grade ore had about 40 percent iron and had to be washed to remove excess sand and rocks. By the late 1950s, the class-one and class-two iron ore deposits were being depleted, and the available ore was reduced. There were millions and millions of tons of lesser iron ore available, called taconite.

On March 25, 1967, the Great Northern began operation of a taconite facility conveyor system. The system consisted of a series of belts handling 3,000 tons per hour (and were weighed as the belt went along) from the unloading dump to stockpile ore to either side of Ore Dock One. The conveyor belt was 2 1/8 miles long.

Taconite is a flint-like2 rock containing 25-30 percent iron distributed in small particles. Processing liberates the ore and forms marble-size pellets, which are then baked in special ovens to harden.

For taconite service the Great Northern added 18 inches to the height of 220 75-ton ore cars. Taconite weighs less than natural ore and the 18 inches is added so that the ore car will hold 75 tons. The ore trains had 200 ore cars and transported to the Allouez yard and back to the taconite facilities.

The pellets, once on Ore Dock One, went into a tripper on a boom, which deposited the pellets into any of the 374 pockets, which have capacities of 275 tons. The tripper rides on rails that span the whole length of the dock. The pellets are eventually loaded from ore pockets down spouts by gravity from the loaded pockets into the holds of ore boats positioned below.

At the end of the Great Northern's last full ore season in 1969, the Great Northern stripped 9.5 million long tons of iron ore and almost 5 million long tons of taconite.

The last boat loaded was the *Joseph H. Thompson* (shown on page 112) on December 19, 1969 (which was also the first taconite-loaded boat at the start of taconite service in 1969).

In 1969, the Great Northern loaded Canadian potash from Ore Dock Four, the first time that commodity was loaded.

Taconite-loaded ore pockets on Ore Dock One await ore carrier. *Author's Collection*

Taconite from the Allouez yard is brought up to Ore Dock One on a conveyor belt. ***Author's Collection***

The Ore Car Tracks have been removed on the left side of Ore Dock One's approach to accommodate for taconite-loading equipment. Pedestal crane and Robbins car shaker are on top of Gantry crane. Northern Pacific ore dock is off on the left side (now demolished). ***Author's Collection***

95

Taconite loading equipment under the Gantry crane ran the length of Ore Dock One on rails. **Author's Collection**

The dumper dumps taconite pellets into one of the pockets on Ore Dock One. **Author's Collection**

Taconite being loaded into pockets. Taconite belt is covered as dumper is dumping taconite far down the deck of Ore Dock One. ***Author's Collection***

C-1 080 steam engines were the workhorse steam engines used on ore dock service. ***Burlington Northern Santa Fe Railway***

This 1500 horsepower EMD SD7, number 559, is pulling empty ore cars off Ore Dock Three. At times, the ore docks had 400 workers operating 24-hours-a-day while shipping was underway. Work would be suspended during periods of rain, sleet, snow or high winds. Work procedures were stated in a work agreement with the Great Northern. *Great Northern Railway Historical Society*

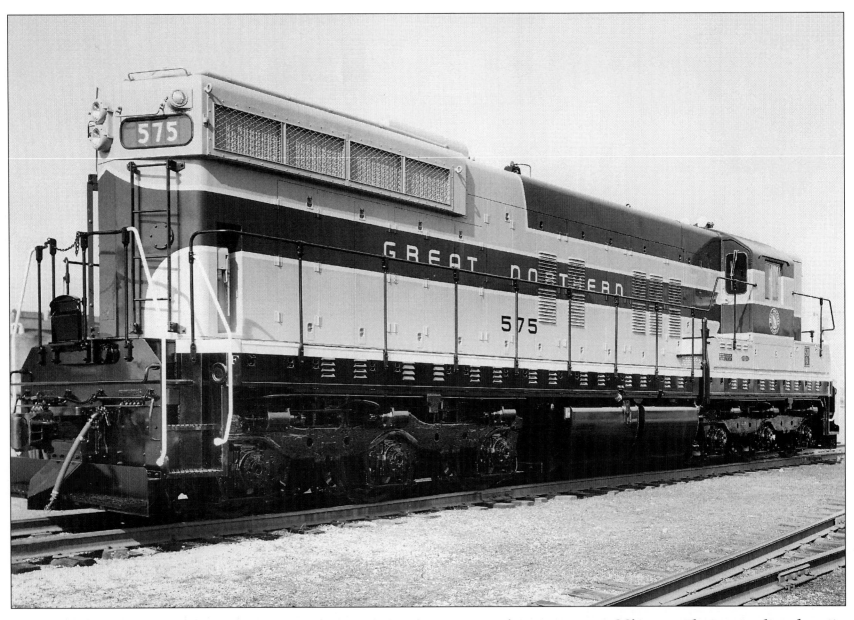

This SD9 with 1,750 horsepower, and SD7 with 1500 horsepower (seen on page 99), were the main diesel units to be used in ore car service on the Great Northern ore docks. ***Burlington Northern Santa Fe Railway***

The first cars used were wood and carried 25-ton loads. The first steel cars used carried 50 tons. The next upgrade was to 75-ton cars as seen with this three-quarter view of 75-ton ore car. All types of cars were 24 feet long and unloaded ore or taconite into two ore pockets on docks. *Great Northern Railway Historical Society*

Inside details of 75-ton ore car. Note doors of bottom open up to dump iron ore. *Great Northern Railway Historical Society*

For taconite service the Great Northern added 18 inches to the height of 220 75-ton ore cars. The Great Northern, from its beginning to its merger with the Burlington Railway, shipped over 977 million tons; over 13 million carloads of iron ore. The ore cars at 24 feet each, if set end to end, would have extended 60,606 miles. *Great Northern Railway Historical Society*

Steam engine pulls twenty-five ton wood ore cars on their way to the Allouez iron ore facilities. The first year the Great Northern shipped ore it totaled 6,359,141 tons. In 1953, the company's most productive year, the Great Northern shipped 32,330,722 tons. *Burlington Northern Santa Fe Railway*

Butting 75-ton ore cars in the Allouez yard waiting to be shoved onto one of the four ore docks. The first steel ore cars were 50 tons. The wood ore cars were 25 tons. An 080 moves into position to distribute ore cars. ***Burlington Northern Santa Fe Railway***

Ore Cars were weighed before going onto the ore docks. ***Burlington Northern Santa Fe Railway***

A C-1 080 steam engine is pushing a shove of ore cars onto Ore Dock Three. A Great Northern employee is standing on top of the ore in an ore car in front of the 080. There was usually a man positioned on top of the ore car in front of the engine and sometimes one in the middle ore car. Hand signals were used to spot the ore cars over the assigned pockets. The engineer would respond to the hand signals from workers on his left view. A fall would mean a 40-foot drop into the bottom of an ore pocket. ***Unknown***

By the 1940s this trapper cart opened trap doors on ore cars. Prior to this, a ratchet system or large wheel opened all cars. **Great Northern Railway Historical Society**

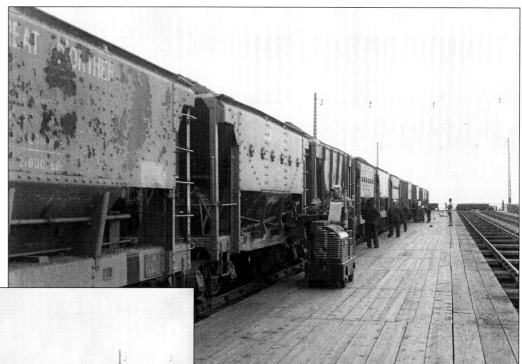

Ore cars correctly spotted over ore pockets. Dock workers prepare to dump a string of cars. **Great Northern Railway Historical Society**

Two dock workers use a wheel to open trap doors on ore cars. The wheel was also used to close trap doors. ***Great Northern Railway Historical Society***

Two-man crew using a bar to turn ratchets that in turn open ore car's dumping doors. The same process is used to close doors. ***Great Northern Railway Historical Society***

Taconite in Ore Dock One pockets awaiting ore boat to be loaded. **Burlington Northern Santa Fe Railway**

Left: Ore Dock One's workman trips a lever, lowering spout above an ore boat. **_Burlington Northern Santa Fe Railway_**

Right: Workman controls flow of ore down spout and into hold of ore boat below. Night loading provided only one small light above ore pocket door. **_1937 photo courtesy of Great Northern Railway Historical Society_**

Photo (circa 1911) from left to right; Ore Dock Four (Concrete and steel) Ore Docks Three, Two and One are all-timber construction. Ore carrier is between end trestles of Ore Docks Two and Three. By 1928 Ore Docks One and Two were converted to concrete and steel construction. *Courtesy Northeast Minnesota Historical Center, University of Duluth. Photo by Louis P. Gallagher*

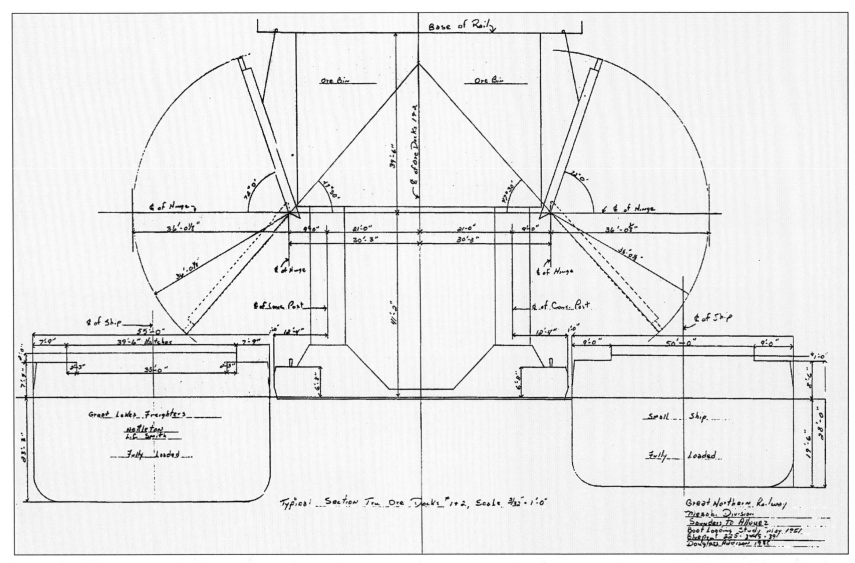

Great Northern blueprint drawing shows how ore boats were loaded. The ore was dumped on top of the ore dock and into the slanted pocket. Ore pocket door was opened after spout had been lowered and ore went into hold of the ore boat. The whaleback ore carriers in the 1890s had varied sizes but the smallest this author found was 320 feet long, 42 feet wide and weighed 2,234 tons. The **Edward Fitzgerald** was 711 feet long and 75 feet wide and tonnage was 13,632 tons. ***Great Northern Railway Blueprint, courtesy Burlington Northern Santa Fe Railway. Drawing by author***

A famed whaleback ore boat built in nearby Superior, Wisconsin, loads first shipment of Mesabi ore at old Ore Dock One in Allouez, destined for Cleveland, Ohio, shown November 11, 1892. The Great Northern purchased the dock in 1899, a catalyst for building the other three docks. **Great Northern Public Relations**

Cigar-shaped whaleback ore boats of 1890 vintage. The ore boat low in the water on the right is the front end of a loaded boat. The boat on the left is the rear end of an unloaded boat. There is one surviving whaleback ore boat, the **Meteor**. At 366 feet long, 45 feet wide and 2,234 tonnage it is currently a museum in Superior, Wisconsin. The Great Northern Railway purchased Ore Dock One from the Duluth, Mississippi River and Northern Railway in 1899. **Unknown**

Workman on barge moving towards Ore Dock One to avoid incoming ore boat carrier. **Burlington Northern Santa Fe Railway**

C-1 080 transforming iron ore to Ore Dock Two. C-1s were the usual steam engines used on ore docks after 1918. Tugboats were used to spot ore boats between the docks. Later, ore boat carriers used jet-propulsion systems to maneuver without the help of tugboats. ***Burlington Northern Santa Fe Railway***

SD unit on Ore Dock Two with a shove of 29 ore cars. Ore Dock Two is 2,100 feet long, 80 feet 6 inches from water to rails on top deck and has 350 ore pockets. Ore Dock One was the largest ore dock at 2,244 feet. ***Burlington Northern Santa Fe Railway***

Four ore boat carriers take up their locations between Ore Docks One and Two. There was 200 feet between the ore docks so ships could load across from each other and still have a tugboat push another ore boat between them. *Burlington Northern Santa Fe Railway*

Two C-1 080 steam engines work to keep up with getting ore cars unloaded to fill the two ore carriers alongside Ore Dock One. The slip occasionally had to be dredged to a depth of 22 feet. *Great Northern Railway Historical Society*

Loading ore in this 1909 photograph at Ore Dock Three is the **C. H. Mc-Cullough, Jr.**—one of 309 boats loaded at Allouez that year. The end trestle has a toilet house on the top deck. **Great Northern Public Relations, Great Northern Railway Historical Society**

Ore Dock Two's Robbins car shaker's clam is on one car to shake ore loose and into ore pocket. Eighty percent of ore cars contained compacted or sticky ore. In the spring and fall the ore would freeze. Steam engines in the Allouez yard produced steam that ran through hoses and finally a rod. The rod was pushed into the five portals on the side of each ore car and taken up on the ore docks. Later, two steam buildings were built to increase ore thawing and allow for longer shipping seasons. The white dots you see on the water are sea gulls. ***Great Northern Public Relations, Great Northern Railway Historical Society***

With anchor down, the ore carrier *Joseph H. Thompson* is awaiting positioning. **Burlington Northern Santa Fe Railway**

The **Edward L. Ryerson** begins taking on taconite from Ore Dock One. **Author's Collection**

Fantail of the **Edward L. Ryerson**. Registry Indiana Harbor. **Author's Collection**

Two dock workers load the ore carrier taking hand signals from the first mate of the **Edward L. Ryerson**, who is in command of loading. **Author's Collection**

Taconite pellets slide down ore spouts into the hold of the **Edward L. Ryerson**. One or two spouts could be used in each hold, as boat holds were either 12 or 24 feet between partitions. Lifting and lowering the spouts helped create an even dump to keep the ore carrier from becoming lopsided. **Author's Collection**

Crew members observe loading of ore. The hold covers are stacked and waiting. Taconite kept the spouts shiny. *Author's Collection*

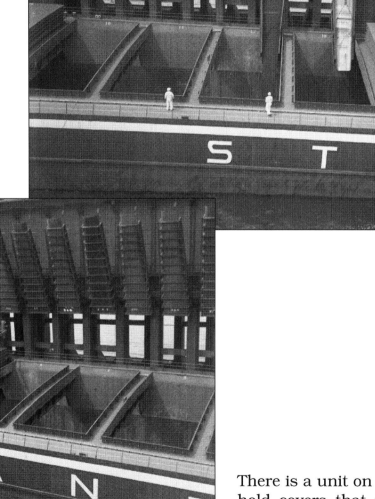

There is a unit on rails over the stacked hold covers that moves on the rail to reach and take hold covers off and on. *Author's Collection*

Bow of **Edward L. Ryerson** and ore being loaded. Two crewmen on the ore dock fender and two crewmen behind the bow deck house show the immense size of Ore Dock One and the **Edward L. Ryerson**. **Author's Collec-**